Advanced Ferroelectric
and
Piezoelectric Materials

With Improved Properties and their Applications

Recommended Titles in Related Topics

Innovative Piezo-Active Composites and Their Structure –
Property Relationships
by James I Roscow, Vitaly Yu Topolov, Christopher R Bowen and
Hamideh Khanbareh
ISBN: 978-981-12-6159-6

100 Years of Ferroelectricity 1921–2021
edited by Julio A Gonzalo, Gines Lifante and Francisco Jaque
ISBN: 978-981-12-4309-7

Essentials of Piezoelectric Energy Harvesting
by Kenji Uchino
ISBN: 978-981-12-3463-7

Mechanics of Piezoelectric Structures
Second Edition
by Jiashi Yang
ISBN: 978-981-12-2679-3

Progress in Advanced Dielectrics
edited by Li Jin
ISBN: 978-981-12-1042-6

Advanced Ferroelectric
and
Piezoelectric Materials

With Improved Properties and their Applications

Ivan A Parinov
Southern Federal University, Russia

Sergey V Zubkov
Southern Federal University, Russia

Alexander S Skaliukh
Southern Federal University, Russia

Valery A Chebanenko
Southern Scientific Center of the Russian Academy of Sciences, Russia

Alexander V Cherpakov
Southern Federal University, Russia

Yuri E Drobotov
Southern Federal University, Russia

World Scientific

NEW JERSEY · LONDON · SINGAPORE · BEIJING · SHANGHAI · HONG KONG · TAIPEI · CHENNAI · TOKYO

Published by

World Scientific Publishing Co. Pte. Ltd.

5 Toh Tuck Link, Singapore 596224

USA office: 27 Warren Street, Suite 401-402, Hackensack, NJ 07601

UK office: 57 Shelton Street, Covent Garden, London WC2H 9HE

Library of Congress Control Number: 2024933454

British Library Cataloguing-in-Publication Data
A catalogue record for this book is available from the British Library.

ADVANCED FERROELECTRIC AND PIEZOELECTRIC MATERIALS
With Improved Properties and their Applications

ISBN 978-981-12-8424-3 (hardcover)
ISBN 978-981-12-8425-0 (ebook for institutions)
ISBN 978-981-12-8426-7 (ebook for individuals)

For any available supplementary material, please visit
https://www.worldscientific.com/worldscibooks/10.1142/13622#t=suppl

Typeset by Stallion Press
Email: enquiries@stallionpress.com

Preface

The book presents some progress and results in the areas of obtaining, converting and storing energy found in the environment, obtained mostly for the last decade by researchers of Piezoelectric Material Science from the well-known Southern Federal University, Rostov-on-Don, Russia. The authors have published a lot of books, devoted to ferroelectrics and piezoelectrics, related materials and their applications [7, 56, 84, 301–318, 358, 481]. This monograph is a new step in this field and is divided into six chapters.

Chapter 1 describes the manufacturing processes and the study of new layered bismuth oxides with the structure of Aurivillius phases, which are ferroelectrics with a high Curie temperature of $T_C > 900°$C. They are characterized by a record high stability of piezoelectric and dielectric parameters in a wide temperature range (70–900 K) and pressures (0.001 Pa–300 MPa). The structural features of these oxides make it possible to widely vary their chemical composition, which leads to significant changes of electrical and magnetic properties. The optimization of their dielectric characteristics by varying their composition and crystal structure is discussed in this chapter.

Chapter 2 studies fractional integro-differentiation operators that are directly involved in the formulation of the dynamical problem of displacement of the ferroelectric domain boundaries. Switching polarization in ferroelectrics is the result of the formation of self-similar structures, leading to fractality of electrical responses. It is considered the general role of fractional analysis operators in the description of these environments and interactions. Some theoretical achievements in the field of such operators are presented, focusing on the Hölder formalism of their smoothness.

Chapter 3 is devoted to the investigation of the physical character-istics of ceramics with inhomogeneous polarization. The problem of longitudinal vibrations of a transverse polarized rod is considered. Then, a mechanical system is studied, consisting of an acoustic path with source and receiver of oscillations in the presence of impedance attenuation, in which piezoceramic elements have uniform polarization. Finally, a problem is considered, close to the previous one, but the piezoceramic transducers of the source and receiver of vibrations have non-uniform polarization.

Chapter 4 presents the state-of-the-art of piezoelectric energy harvesting developed over the last decade. First, the obtained results from the experimental studies and the modeling of cantilever-type piezoelectric generators (PEGs) with proof masses and based on the porous piezoceramics are discussed. Then, the problems of nonlinearity and expanding the frequency broadband for efficient monostable, bistable, tristable and multistable energy harvesters are considered. The PEGs with constructive L-shaped elements are also discussed. Finally, the results for stack-type piezoelectric generators and some medical piezoelectric applications are examined.

Chapter 5 presents expanded test and finite element models (FEMs) for cantilever-type and axial-type PEGs with active ele-ments. The results cover various structural and electric schemes of the PEGs with proof mass, bimorph and cylindrical piezoelectric elements, and excitation loads. Based on the FEMs, modal and harmonic analyses are performed. Resonance frequencies, output voltage and power are calculated.

Chapter 6 reviews some results obtained over the last five years from modeling the vibration of devices from piezoactive materi-als, including the five following effects: piezoelectric, flexoelectric, pyroelectric, piezomagnetic and flexomagnetic. Analytical, semi-analytical and FE analysis are the main methods for modeling vibra-tions of devices, operating on these effects. Various approaches are used, such as the Hamilton principle, Euler–Bernoulli or Timoshenko beam theories, shear strain theories of various orders, nonlocal strain gradient elasticity and couple stress theory.

This book is intended for a wide range of students, engineers and specialists interested in and participating in the research and development of the modern problems of ferroelectric and piezoelectric Material Science.[1]

[1]Research was financially supported by Russian Science Foundation, grant No. 21-19-00423 (Chapters 1, 3, 6) and Ministry of Science and Higher Education of the Russian Federation (State task in the field of scientific activity, project No. FENW-2023-0012) (Chapters 2, 4, 5).

Contents

List of Tables

List of Figures

Chapter 1

Crystal Structure and Electrophysical Characteristics of Layered Perovskite-like Bismuth Oxides

1.1. Introduction

In the development of the theory of ferroelectricity, which is now an important area of solid-state physics, an exceptional role has been played by the discovery by B. M. Wool in 1944 (Physical Institute of the USSR Academy of Sciences) of the ferroelectric properties of one of the most remarkable structural analogues of perovskite, barium titanate. It marked the beginning of a broad search for new ferroelectrics, primarily with perovskite and perovskite-like structures.

Already in 1949, the Swedish chemist B. Aurivillius (Bengt Aurivillius), studying the Bi_2O_3–TiO_2 system, established the formation of a three-layer oxide $Bi_4Ti_3O_{12}$ with a perovskite-like structure [22].

In 1950, V. Aurivillius studied the crystal structure of tantalite niobates of the type $ABi_2Nb_2O_9$, $ABi_2Ta_2O_9$ ($A = Ca, Sr, Pb, Ba$), Bi_3NbTiO_9 and $Bi_4Ti_3O_{12}$. In particular, he showed that, at room temperature, the unit cells of $CaBi_2Nb_2O_9$, $CaBi_2Ta_2O_9$, Bi_3NbTiO_9 and $PbBi_2Nb_2O_9$ had rhombic symmetry and consisted of a perovskite layer perpendicular to [001] and $Bi_2O_2^{2+}$ layers [23].

Then, within two years, he obtained several more oxides with a similar structure [24]. However, at the first stage, V. Aurivillius

1

limited himself by studying only the structure of the obtained compounds. At about the same time, G. A. Smolensky and N. V. Kozhevnikova pointed out that structures that have ions of small size and large charge (for example, Ti^{4+}, Nb^{5+}, Ta^{5+}) in oxygen octahedra, coupled through angles, forming continuous chains of "oxygen–metal–oxygen", are favorable for the occurrence of ferroelectricity [369].

A decade later, G. A. Smolensky, V. A. Isupov and A. I. Agranovskaya discovered that the compound Bi_2PbNbO_9, belonging to this class of compounds, is a new ferroelectric with a Curie temperature of $T_C = 520°C$ [368].

In 1960, I. G. Ismailzade also studied the crystal structure of polycrystalline samples $CaBi_2Nb_2O_9$, $CaBi_2Ta_2O_9$, $PbBi_2Ta_2O_9$ and Bi_3NbTiO_9, investigating the nature of the spontaneous polarization of these compounds [168].

Subsequently, V. Aurivillius and E. Subbarao [25, 397, 398, 400] obtained and studied several dozen more phases of perovskite-like bismuth oxides, which later received their proper name, Aurivillius phases (APs).

Immediately after this, intensive research began on other compounds of this type. As a result, an extensive class of new ferroelectrics was discovered, many of which were distinguished by high Curie temperatures and high stability of ferroelectric properties. Therefore, interest in APs is fueled by the prospects of their application in high-temperature piezoelectric sensors, as well as in highly reliable memory devices. In addition to ferroelectric properties, compounds of this class also exhibit ionic conductivity.

To date, hundreds of Aurivillius phases of various compositions are known, and thousands of studies have been published on their properties and structures.

Information about these compounds, obtained before 1985, was systematized in a monograph edited by G. A. Smolensky [367]. This chapter summarizes the main results of studies of the structural and crystal chemical aspects of APs obtained over the past decades.

1.2. Structure and Existing Conditions of Aurivillius Phases

There are several perovskite-like oxides with a layered microstructure: Dion-Jacobson phases $(A_m B_m O_{3m+2})$, Ruddlesden-Popper phases $(A_{m+1} B_m O_{3m+1})$ and APs $(A_{m-1} Bi_2 B_m O_{3m+3})$, where m is the number of perovskite layers (see Figure 1.1).

The chemical composition of the APs is described by the general formula $A_{m-1} Bi_2 B_m O_{3m+3}$, where A are the ions in positions with a cuboctahedral environment, which can be one-, two- and trivalent cations of large radius (Na^+, K^+, Ca^{2+}, Sr^{2+} [511], Ba^{2+}, Pb^{2+}, Bi^{3+}, Gd^{3+} [507], Y^{3+} [512], Ln^{3+} (lanthanides) and Ac, Th [439, 508] (actinides), while B-positions inside the oxygen octahedra are occupied by highly charged ($\geq 3+$) cations with small radii (Cr^{3+}, Ga^{3+}, Mn^{4+}, Fe^{3+}, Co^{3+}, Ni^{3+} [441], Ti^{4+}, Ta^{4+} [513], Nb^{5+}, W^{6+} [440], Mo^{6+}, Ir^{4+}, etc.).

Aurivillius phases $A_{m-1} Bi_2 B_m O_{3m+3}$ form a tetragonal lattice in the high-temperature paraelectric phase. They have deformation components in the ferroelectric state (and, as a result, spontaneous polarization) along c-axis and along the rhombic axis b, dividing the

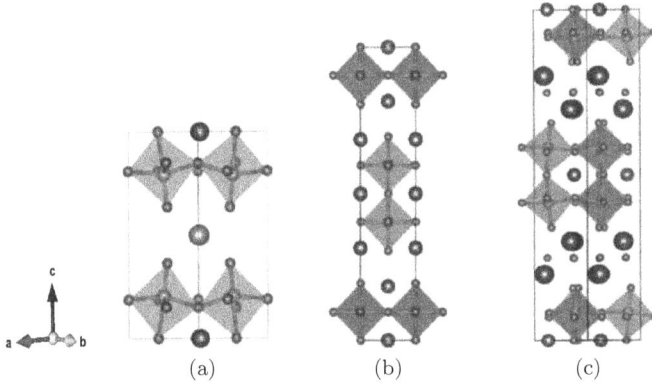

Fig. 1.1. Perovskite-based intergrowth structures: (a) Dion-Jacobson, (b) Ruddlesden-Popper, (c) Aurivillius phases.

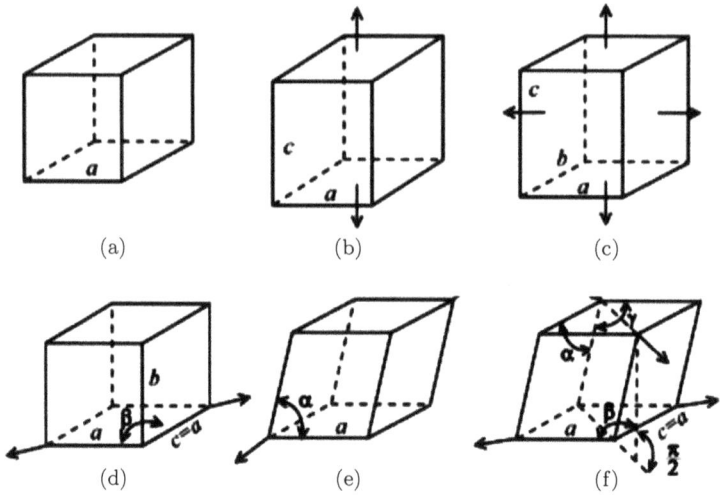

Fig. 1.2. Cells of perovskite compounds: (a) cubic; (b) tetragonal; (c) rhombic I; (d) rhombic II; (e) rhombohedral and (e) triclinic.

angle between a-axis and b-axis in half, and APs can be monoclinic, rhombic, tetragonal (see Figure 1.2).

The perovskite-like layer of the AP unit cell has a structural fragment in which six $X = O$ oxygen ions form a crystallographic polyhedron in the form of an octahedron around the smaller B-cation, and eight large A-cations can form a cube. Twelve oxygen ions form a cuboctahedron. For each oxygen ion $X = O$, the nearest neighbors will be both cation A and cation B, located in the form of a tetragonal bipyramid and having different sizes and properties. Four large cations A form the base of a bipyramid with sides equal to the unit cell parameter a. Two cations B ($R_B < R_A$, where R_A and R_B are the ionic radii of the cations) are located perpendicular to the center of the base of the bipyramid on both sides at distances $a/2$.

The value of m is determined by the number of $[A_{m-1}B_m O_{3m+1}]^{2-}$ perovskite layers, located between the $[Bi_2O_2]^{2+}$ fluorite-like layers, and can take integer or half-integer values in the range of $m = 1 - 5$. If m is a half-integer number, then in the lattice there is an alternation of perovskite layers with m differing by 1. If the perovskite-like layer of minimum thickness $m(P = 1)$ is denoted by P, and the bismuth-oxygen layer by B, then the layered compounds can be described as $BPmBPm$. Neighboring layers of BPm are shifted relative to each other in the [110] direction by half of the diagonal of the base of the tetraganal unit cell, which entails a doubling of the number of BPm layers in unit cell $2(BPm)$, that is doubling the c-period. If perovskite-like layers of different thicknesses (m_1m_2) alternate in a layered joint, then the average layer thickness m can be a fractional number. For example, if $m_2 = m_1 + 1$, then the average layer thickness is $m_1 + 1/2$ (see Figure 1.3). Since adjacent perovskite-like layers are already shifted relative to each other, the difference in the thicknesses of adjacent layers does not lead to an additional doubling of the c-period [173].

The AP structure above the Curie point T_C is tetragonal and belongs to the space group $I4/mmm$. The type of space group below the Curie point T_C depends on the value of m. For odd m, the space group of the ferroelectric phase is $B2cb$ or $Pca2_1$, for even m, $A2_1am$, $Fmm2$, $Fmmm$ and for half-integer, $Cmm2$ or $I2cm$.

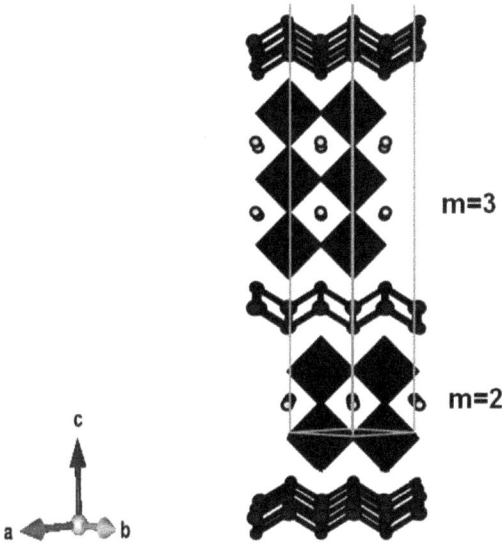

Fig. 1.3.   Crystal structure of AP compound $Bi_2A_{m-1}B_mO_{3m+3}$ $(m = 2.5)$.

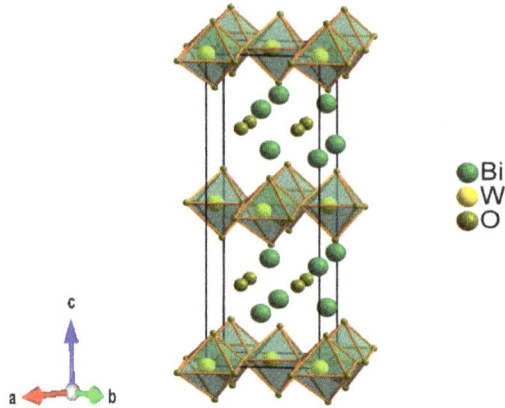

Fig. 1.4. Crystal structure of the AP compound Bi_2WO_6 with $m = 1$.

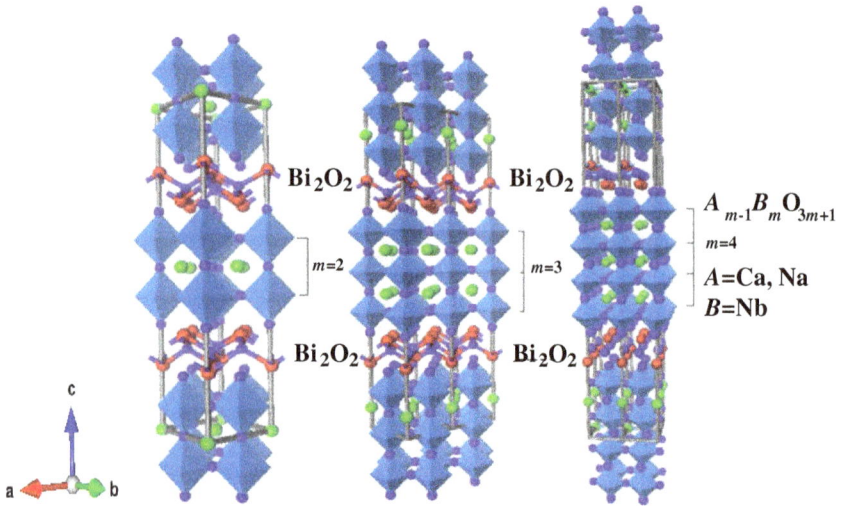

Fig. 1.5. Crystal structures of the AP compounds with $m = 2, 3, 4$.

Figures 1.4 and 1.5 show AP structures with $m = 1$ and $m = 2, 3, 4$, respectively. Positions A and B can be occupied by the same or several different atoms. The substitution of atoms in positions A and B has a significant effect on the electrophysical characteristics of the APs. There are large changes in the values of dielectric constants and conductivity, and the Curie temperature T_C can also vary over a wide range.

In a lot of works, a detailed crystallochemical analysis of ferroelectrics from different APs was carried out. So, in Refs. [128, 336, 396] the conditions for the existence of these compounds were considered, which can be expressed by the following formula:

$$t_1 < t = [(R_A + R_O)/\sqrt{2}(R_B + R_O)] < t_2, \qquad (1.1)$$

where R_A, R_B are the radii of cations, R_O is the radius of oxygen anions. The value of t is called the Goldschmidt tolerance factor. The boundary values of the tolerance factor, which determine the possibility of the existence of a compound, related to AP, were defined as $t_1 = 0.870$ and $t_2 = 0.985$. A more detailed study of the patterns of changes in the Curie temperature T_C of the APs, depending on such parameters as the radii and electronegativity of the A- and B-ions, as well as on the cell parameters, was carried out in Ref. [105]. Some anomalies in the properties of layered ferroelectrics were also considered for compounds $A_{m-1}Bi_2B_mO_{3m+3}$[243]. In particular, it has been shown that Bi_2O_2 layers exert a tightening effect on the layered structure of these compounds, but the strength of this effect decreased as the number of perovskite layers increased. In addition, it was shown that the Curie temperatures T_C pass through a maximum value at increasing distortion of the pseudoperovskite cell in the perovskite-like layer, and the position of the maximum value changes with the number of perovskite layers m.

The study of structural features of the APs $Bi_{m+1}Fe_{m-3}Ti_3O_{3m+3}$ in the Bi_2O_3–Fe_2O_3–TiO_2 system showed the presence of sharp structural changes with an increase in the thickness of the perovskite-like layer to $\sim$2 nm (for $m \approx 5$). As the thickness of the perovskite-like block increase to $\sim$3.7 nm (for $m \approx 9$), APs become unstable. This correlates with the equalization of the effective charges of octahedrally surrounded ions in two structurally nonequivalent positions of the perovskite-like block. The APs instability also correlate with a decrease in order in the distribution of Fe^{3+} and Ti^{4+} ions to a completely disordered distribution at $m \approx 9$ between the inner and outer layers of the perovskite-like block and the approach of the average thickness of the perovskite-like layer to the corresponding parameter of the compound $BiFeO_3$ [174, 250, 251].

In ferroelectric compounds of the perovskite family, as well as layered materials such as APs, the properties are closely related to small distortions of an ideal, highly symmetrical structure. In this regard, obtaining and analyzing information about such distortions can provide a key to understanding the nature of ferroelectricity in these materials and developing ways to obtain new materials with desired properties. Over the past decades, the structures of APs with $m = 2$ and $m = 4$ have been intensively studied in order to reveal the patterns of their structures. Almost all compounds, namely 17, of these series, described in the literature at room temperature, have an orthorhombic cell with the space group $A2_1am$ and close cell parameters. The exception was related to $BaBi_2Ta_2O_9$, refined in the highly symmetric space group $I4/mmm$; its cell parameters were equal to $a = 3.92650(8)$ Å and $c = 25.5866(8)$ Å [359].

In a lot of works, a series of comparative studies have been carried out in APs with different atoms in the A-positions [37, 41, 42, 171, 401, 414, 464]. In the process of these studies, some features of the structure were revealed, which, to some extent, are characteristic of all APs. Thus, the atoms, located in the A-positions and B-positions, were displaced along the c-axis in the direction toward the oxygen atoms, lying in the fluorite-like layers. As one moved away from the fluorite-like layers, this shift became smaller. This expansion of the cationic sublattice was combined with contraction of the anionic sublattice, leading to the formation of short B–O bonds. The shift of cations in A-position also took place along the polar a-axis.

For example, based on the powder diffraction patterns, the structures of three APs of the $ABi_2Ta_2O_9$ series (where $A = Ca$, Sr, Ba [359]) were refined; they are shown in Figure 1.6. The study made it possible to trace the influence of the A-cation value on the degree of structural distortions. The figure clearly shows how the distortion of TaO_6 octahedra increases as the size of the cation in the A-position decreases.

A similar comparative study for $Bi_{2.5}Na_{0.5}Nb_2O_9$ and $Bi_{2.5}K_{0.5}$ Nb_2O_9 was described in Ref. [41]. Later, two other compounds of

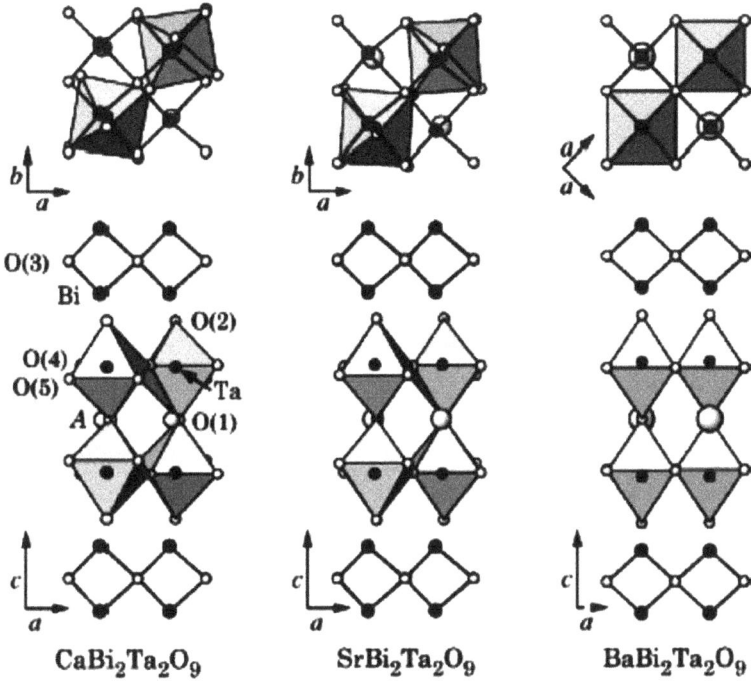

Fig. 1.6. Crystal structure of $ABi_2Ta_2O_9$, where A = Ca, Sr, Ba.

this series were studied: $Bi_{2.5}Na_{0.5}Ta_2O_9$ [42] and $Bi_{2.5}Ag_{0.5}Nb_2O_9$ [464]. Here, the same regularity was observed as in the previous case, and rhombic distortions increased in the sequence: $A =$ K−Ag−Na.

In relation to the rotation of octahedra, there is no obvious dependence on the size of the cation A, and in $Bi_{2.5}Ag_{0.5}Nb_2O_9$ the rotation of octahedra along c-axis turns out to be greater than in $Bi_{2.5}Na_{0.5}Nb_2O_9$ and $Bi_{2.5}K_{0.5}Nb_2O_9$, which may be due to the difference in the electronic structure of these atoms. The displacement of Nb atoms along the c-axis with respect to the centers of oxygen octahedra is the largest for the K-containing AP compound and decreases with increasing an ion radius of A. In comparison with the isostructural Nb analog, the structure of $Bi_{2.5}Na_{0.5}Ta_2O_9$ is less distorted, and calculations using the valence sum method show that the Ta-O bonds are somewhat stronger than the Nb-O bonds.

Similar studies were carried out for a number of APs with $m = 4$, such as $BaBi_4Ti_4O_{15}$, $CaBi_4Ti_4O_{15}$, $SrBi_4Ti_4O_{15}$, $PbBi_4Ti_4O_{15}$ [401, 414], and also with $m = 5$, such as $Ba_2Bi_4Ti_5O_{18}$, $Ca_2Bi_4Ti_5O_{18}$, $Sr_2Bi_4Ti_5O_{18}$, $Pb_2Bi_4Ti_5O_{18}$ [471]. A comparative study of the structures of $BaBi_4Ti_4O_{15}$ and $CaBi_4Ti_4O_{15}$, which represent two extreme cases with respect to the radius of the cation in A-position, was carried out in Ref. [118]. In terms of layer packing along c-axis, both structures are similar and exhibit features typical for all APs. The most obvious differences in the structures are due to the degree of their distortion, which is much higher for the Ca-containing compound. In $CaBi_4Ti_4O_{15}$, the oxygen octahedra are rotated around a- and c-axes alternately in different directions, while in $BaBi_4Ti_4O_{15}$ these distortions are very small, which makes it possible to refine this structure in more highly symmetric space groups, such as $F2mm$ (see Figure 1.7). $BaBi_4Ti_4O_{15}$ also exhibits partial substitution of Bi atoms by Ba atoms in fluorite-like layers. The relaxor properties of the compound are explained in the work by the fact that the positions of the Bi and Ba atoms in the perovskite layers differ significantly, which leads to the formation of microregions with varying degrees of structural distortions.

Structural distortions in the AP series $Ba_2Bi_4Ti_5O_{18}$, $Ca_2Bi_4Ti_5O_{18}$, $Sr_2Bi_4Ti_5O_{18}$ and $Pb_2Bi_4Ti_5O_{18}$ with $m = 5$ also follow general trends, and the degree of structure "crumpling" increases from $Ba_2Bi_4Ti_5O_{18}$ to $Ca_2Bi_4Ti_5O_{18}$ [171, 401, 414]. A comparison of the structures of $BaBi_4Ti_4O_{15}$, $SrBi_4Ti_4O_{15}$ and $PbBi_4Ti_4O_{15}$ [401] revealed strong differences in the coordination environment of Bi atoms, localized in perovskite layers. If the Bi atoms in $BaBi_4Ti_4O_{15}$ are coordinated by 12 oxygen atoms, then the coordination numbers for $SrBi_4Ti_4O_{15}$ and $PbBi_4Ti_4O_{15}$ are equal to 10 and 11, respectively.

The influence of the number of layers on structural distortions was studied using the examples of a number of compounds $Bi_{2.5}Na_{0.5}Ta_2O_9$ ($m = 2$), $Bi_{2.5}Na_{1.5}Nb_3O_{12}$ ($m = 3$) and $Bi_{2.5}Na_{2.5}Nb_4O_{15}$ ($m = 4$) [171]. When the number of layers increases, the parameters a and b increase, which contribute to the growth of stress in the structure. Electron microscopy data also indicate that

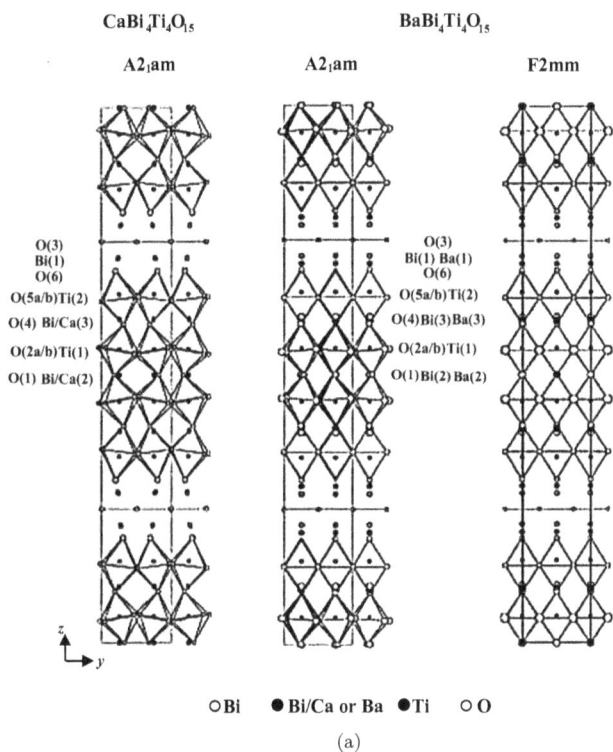

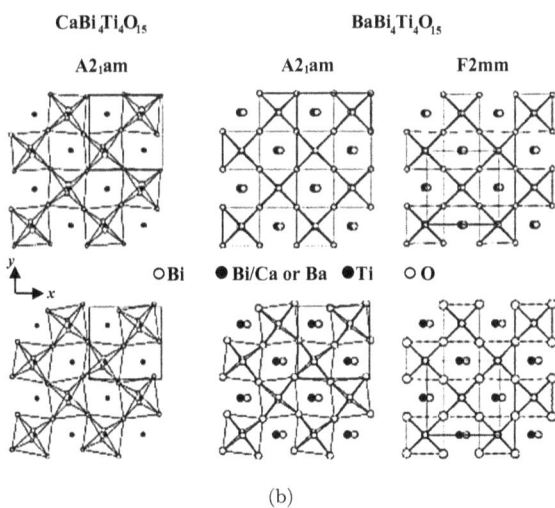

$CaBi_4Ti_4O_{15}$ $BaBi_4Ti_4O_{15}$

A2$_1$am   A2$_1$am   F2mm

O(3)
Bi(1)
O(6)
O(5a/b)Ti(2)
O(4) Bi/Ca(3)
O(2a/b)Ti(1)
O(1) Bi/Ca(2)

O(3)
Bi(1) Ba(1)
O(6)
O(5a/b)Ti(2)
O(4)Bi(3)Ba(3)
O(2a/b)Ti(1)
O(1)Bi(2) Ba(2)

○ Bi   ● Bi/Ca or Ba   ● Ti   ○ O

(a)

$CaBi_4Ti_4O_{15}$ $BaBi_4Ti_4O_{15}$

A2$_1$am   A2$_1$am   F2mm

○ Bi   ● Bi/Ca or Ba   ● Ti   ○ O

(b)

Fig. 1.7. Crystal structure of $BaBi_4Ti_4O_{15}$ and $CaBi_4Ti_4O_{15}$ in the projection on the plane (011) (a) and (110) (b).

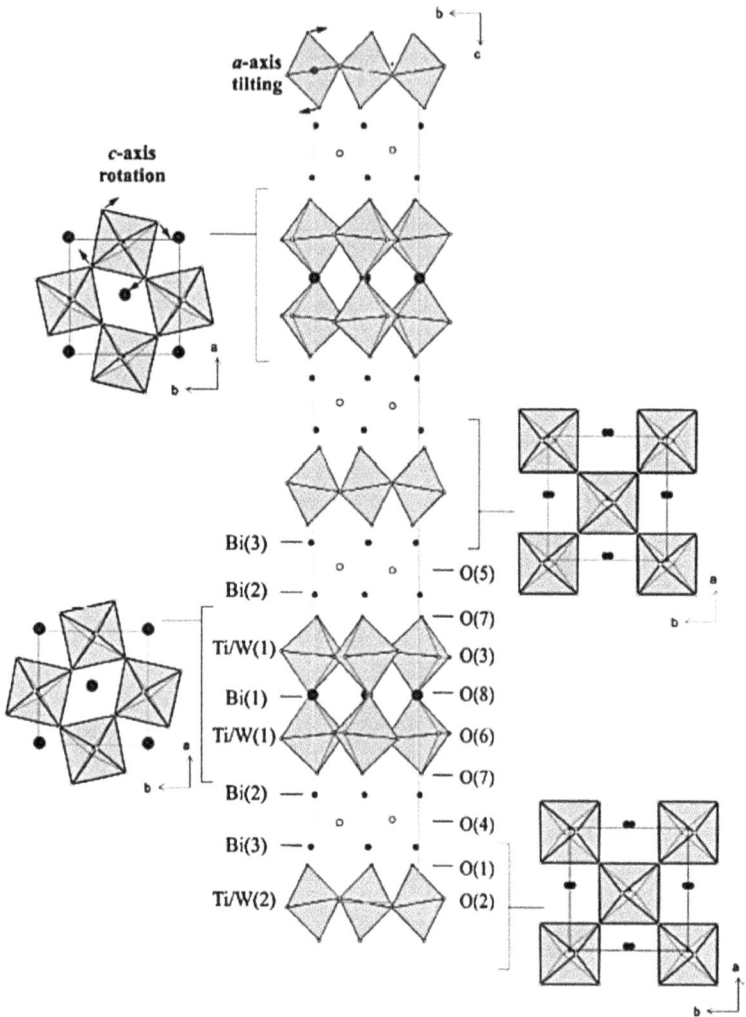

Fig. 1.8. Crystal structure of $Bi_5Ti_{1.5}W_{1.5}O_{15}$.

an increase in the number of layers leads to an increase in the number of defects in the crystal packing.

A detailed analysis of the structural distortions of mixed-layer APs with $m = 1.5$ was carried out by X-ray powder diffraction, using the example of $Bi_5Ti_{1.5}W_{1.5}O_{15}$ [413]. To determine the true parameters of the cell, additional data, obtained by electron diffraction, were used. As a result, the cell was defined as rhombic $I2cm$, with the parameters: $a = 5.4092(3)$ Å, $b = 5.3843(3)$ Å and $c = 41.529(3)$ Å. Based on the complete determination of the structure, carried out by the Rietveld method, three types of structural distortions, characteristic for this structural type, were analyzed, namely: (i) rotation of oxygen octahedra around a-axis; (ii) rotation of oxygen octahedra around b-axis; (iii) displacement of atoms along the polar c-axis.

It has been found that the distinguishing feature of the $Bi_5Ti_{1.5}W_{1.5}O_{15}$ structure is the disappearance of octahedron rotations in single perovskite layers, while they remain in double layers, as shown in Figure 1.8. Moreover, changes in the distances between cations, located at the centers of oxygen octahedra and their oxygen environment, were observed.

1.3. Study of Electrophysical and Magnetic Properties of Aps

APs have a number of electrophysical characteristics that distinguish them from other compounds. This is primarily the existence of a ferroelectric state up to high temperatures ($1000°C$), slight fatigue of repolarization processes and high dielectric constant, which makes it possible to obtain thin films that are promising as nonvolatile memory elements (ferroelectric random access memory (FRAM)).

In structures with an oxygen octahedron and ions with small sizes and large charges at their centers (for example, Ti^{4+}, Nb^{5+}, Ta^{5+}), the appearance of spontaneous polarization is often characteristic. In APs, this spontaneous polarization, being the result of various ways of simultaneous rotation of oxygen octahedra relative to each other and displacements of ions, located in B-position of perovskite cell, has a main component in the [100] direction of the a-axis of perovskite-like layers. Thus, the polarization behavior will differ from that in perovskite ferroelectrics with direct polarization in the direction of the c-axis of the perovskite structure [90].

1.3.1.   *Effect of sintering temperature on dielectric properties of APs*

In the work [441], it has been stated that the maximum dielectric constant of $Sr_{0.5}Bi_{4.5}Co_{0.5}Ti_{3.5}O_{15}$ reached values of 1700, while for $Sr_{0.5}Bi_{4.5}Ni_{0.5}Ti_{3.5}O_{15}$ this value was 4 times less (near 480), see Figure 1.9, despite the fact that changes in the composition were

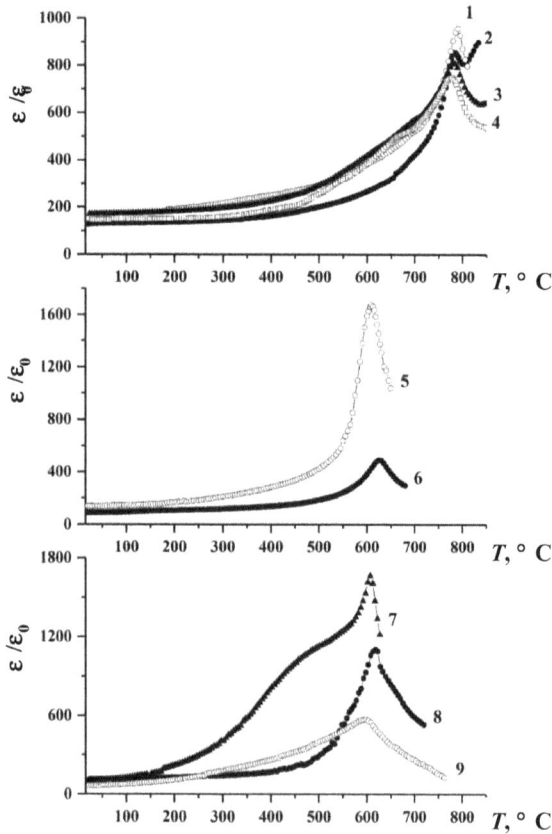

Fig. 1.9. Dependencies $\varepsilon/\varepsilon_0(t)$ at 200 kHz for (1) $Ca_{0.5}Bi_{4.5}Co_{0.5}Ti_{3.5}O_{15}$, (2) $Ca_{0.5}Bi_{4.5}Ni_{0.5}Ti_{3.5}O_{15}$, (3) $Ca_{0.5}Bi_{4.5}Ga_{0.5}Ti_{3.5}O_{15}$, (4) $Ca_{0.5}Bi_{4.5}Fe_{0.5}Ti_{3.5}O_{15}$, (5) $Sr_{0.5}Bi_{4.5}Co_{0.5}Ti_{3.5}O_{15}$, (6) $Sr_{0.5}Bi_{4.5}Ni_{0.5}Ti_{3.5}O_{15}$, (7) $Pb_{0.5}Bi_{4.5}Cr_{0.5}Ti_{3.5}O_{15}$, (8) $Pb_{0.5}Bi_{4.5}Ga_{0.5}Ti_{3.5}O_{15}$, (9) $Pb_{0.5}Bi_{4.5}Mn_{0.5}Ti_{3.5}O_{15}$.

reduced to the replacement of Co with Ni, which was close in ionic radius.

In order to understand the reasons for such significant differences, an additional study of the microstructure of the samples was carried out by using the method of scanning electron microscopy (SEM). As shown in a lot of works [183,186,282], the dielectric properties of APs can be significantly affected by the microstructural characteristics of samples, such as the density, grain size and porosity of the ceramic. According to SEM data, it was found that the $Sr_{0.5}Bi_{4.5}Co_{0.5}Ti_{3.5}O_{15}$ sample is characterized by a significantly lower porosity and larger grain sizes compared to $Sr_{0.5}Bi_{4.5}Ni_{0.5}Ti_{3.5}O_{15}$, despite identical synthesis conditions.

When studying the effect of doping with Nd-ions on the piezo-electric properties of the $Bi_{3-x}Nd_xTi_3O_{12}$ compound, it was found that during sintering at temperatures of $1000°C$ and $1050°C$, the dielectric constants ($\varepsilon'1000$ and $\varepsilon'1050$) of the same samples differed by almost two times (see Table 1.1).

An analysis of the statistical data, obtained from studying the microstructure of the $Bi_{3-x}Nd_xTi_3O_{12}$ samples ($x = 0.1, 0.3, 0.5, 0.7$) by using SEM over the range of crystallite sizes (see Figure 1.10) led to the conclusion that their scatter in both cases followed a standard log-normal distribution with an average size of $\sim 1.1\,\mu m$ (for $T = 1000°C$) and $\sim 2.5\,\mu m$ ($T = 1050°C$).

Significant grain growth was noted with an increase in the sintering temperature. It can be seen that 93% of the sizes of all grains of ceramics, manufactured at $T = 1050°C$, were within

Table 1.1. Values of t-tolerance factor, Curie temperature T_C, relative permittivity $\varepsilon/\varepsilon_0$ and activation energy $E_{1,2}$ for $Bi_{4-x}Nd_xTi_3O_{12}$ ($x = 0.1, 0.3, 0.5, 0.7$) ceramics sintered at $1000°C$ and $1050°C$.

Compound	t-factor	$T_{C1000}/$ T_{C1050}	$\varepsilon'1000/$ $\varepsilon'1050$	E_1/E_2, eV at $1000°C$	E_1/E_2, eV at $1050°C$
$Bi_{3.9}Nd_{0.1}Ti_3O_{12}$	0.978	695/680	1250/5000	0.56/0.04	0.50/0.03
$Bi_{3.7}Nd_{0.3}Ti_3O_{12}$	0.976	655/650	2805/3619	0.55/0.05	0.56/0.02
$Bi_{3.5}Nd_{0.5}Ti_3O_{12}$	0.974	610/645	2300/4800	0.54/0.03	0.59/0.05
$Bi_{3.3}Nd_{0.7}Ti_3O_{12}$	0.971	580/600	1600/3000	0.56/0.02	0.58/0.06

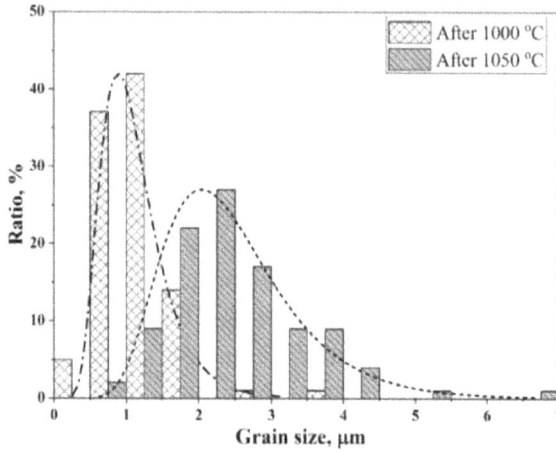

Fig. 1.10. Grain size distribution in $Bi_{3.5}Nd_{0.5}Ti_3O_{12}$ ceramics after synthesis at $T = 1000°C$ and $T = 1050°C$.

1.5–4 μm, while for ceramics manufactured at $T = 1000°$ C, this range was only 1–2 μm.

1.3.2. *Effect of doping on the electrophysical characteristics of APs*

A large selection of cations that can replace the A- and B-positions in the perovskite layers of the APs' crystal structure makes it possible to vary the electrophysical characteristics of the APs over a wide range. A large number of studies have been devoted to the modification of APs' properties by doping cations of various types [90, 118, 471, 485].

In Ref. [471], three series of solid solutions of composition $Bi_{2-x}Te_xSrNb_{2-x}B_xO_9$ $(B = Zr, Hf)$ with $x \leq 0.5$ and $Bi_{2-x}Te_x$ $Sr_{1-x}K_xNb_2O_9$ with $0 \leq x \leq 0.25$ were studied. It was found that the values of the Curie temperature T_C decreased approximately linearly with increasing the parameter $\gamma = \sqrt[3]{abc}$, that is, the larger the sizes of pseudooctahedrons, the lower the Curie temperatures T_C of the studied compounds. It was also found that the Curie temperatures T_C varied approximately linearly with the value of the electronegativity of the cations in the 24th crystallographic B-position. Figure 1.11 shows these dependencies.

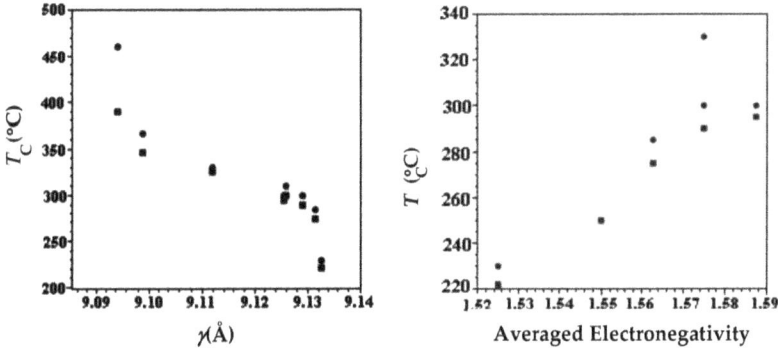

Fig. 1.11. Dependences of T_C on the parameter γ and the average value of the electronegativity of cations in B-position.

The results of the effect of doping on the crystal structure, microstructure and dielectric and electrical properties of $SrBi_2Nb_2O_9$ (SBN)-layered structures were presented in Ref. [471]. The substitution of Ca^{2+} and Ba^{2+} ions in A-positions for Sr^{2+} ions and up to 30 at. % in B-positions (Nb^{5+} for V^{5+}) was studied. It was found that the crystal lattice constants and the dielectric and electrical properties of SBN ceramics significantly depended on the type and number of doping atoms. It has been established that doping with vanadium has a significant effect on the dielectric and ferroelectric properties of doped structures [467–470, 472]. In particular, the residual polarization of the SBN ferroelectric when doped with 10 at. % of vanadium increased from $\sim$2.8 $\mu C/cm^2$ to $\sim$8 $\mu C/cm^2$ and the coercive field decreased by $\sim$63 kV/cm down to $\sim$50 kV/cm. Figure 1.12 shows the effect of doping the SBN system with Ca/Ba- and V-atoms. We can draw unambiguous conclusions that doping the initial composition with Ca- and V-ions increases the Curie temperature T_C when the degree of doping increases. In contrast, doping the initial composition with Ba-ions lowers the Curie temperature T_C, and in this case, the permittivity at T_C almost doubles for 10% of V compared with SBN.

Substituting ions can affect several factors, such as ionic sizes, correlation coefficients, ionic polarization, etc. Isotropic perovskite ferroelectrics doped with large ions at the A-positions and/or small

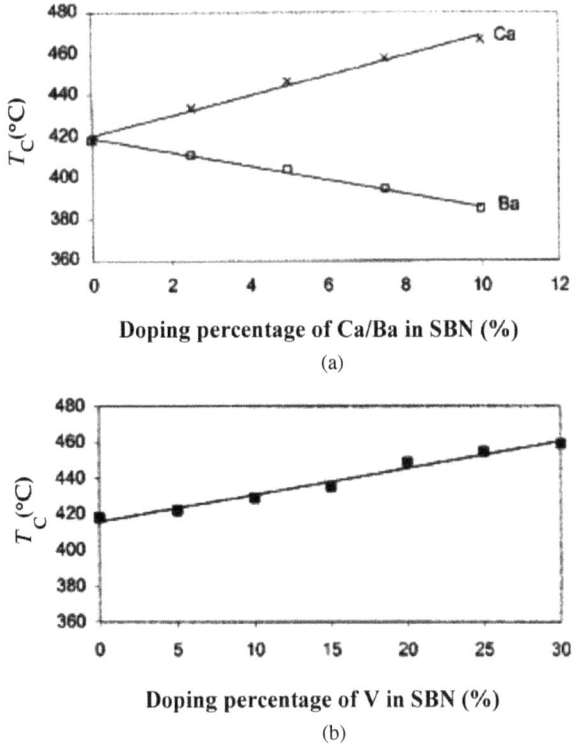

Fig. 1.12. Change in the Curie temperature T_C depending on the doping of SBN with Ca/Ba (a) and V (b) atoms.

ions at the B-positions typically have a high Curie temperature T_C. For example, the Curie temperature T_C in $(Ba, Sr)TiO_3$ varies linearly with the Ba/Sr ratio, and a low Zr/Ti ratio in $Pb(Zr, Ti)O_3$ results in an increased Curie temperature T_C. This is due to the high mobility of ions in B-positions at the center of the perovskite cell.

Further, Pb^{2+} and Bi^{2+} usually lead to a relatively higher Curie temperature T_C, since the upper 6s electron shell is completely filled [469] and can contribute to polarization. Thus, Pb^{2+} and Ti^{4+} in $PbTiO_3$ have a higher mobility than Ba^{2+} and Ti^{4+} in $BaTiO_3$ [108] and a higher Curie temperature T_C than $BaTiO_3$, although the Ba^{2+} ion ($\sim$160 pm) is larger than the Pb^{2+} ion ($\sim$132 pm). In layered perovskites, the crystal structure also cannot freely change upon doping, since doping is limited by the $[Bi_2O_2]^{2+}$ interlayer. When Sr^{2+} ($\sim$143 pm) in A-position is replaced by a large Ba^{2+} ion ($\sim$160 pm), the Curie temperature T_C decreases. When a small Ca^{2+} ion ($\sim$136 pm) replaces the Sr^{2+} ion, the Curie temperature T_C rises. This can be explained by the fact that the introduction of a large ion into A-position requires more space despite the increase in cell parameters. As a result, the mobility of the oxygen octahedron ions decreases and the Curie temperature T_C decreases. Thus, according to Ref. [175], the Curie temperature T_C for CBN and BBN compounds is equal to $\sim$620°C and $\sim$200°C [403], respectively. Such results may explain the fact that the Curie temperature is associated with ionic polarization, which in turn is associated with structural distortions. The larger difference between the parameters a and b for CBN compared to SBN and BBN gives a higher Curie temperature T_C [76, 79, 107, 172, 283, 334, 360, 361, 406, 436, 456, 462, 496]. With this purpose, for example, the authors of Refs. [485, 496] synthesized and studied ceramics of the composition $(Bi_3TiNbO_9)_x(SrBi_2Ta_2O_9)_{1-x}$. These ceramics, depending on $x < 0.8$, had transition temperatures in the range of 330°C $< T <$ 920°C. It was found that the transition temperatures increased linearly with x. Differences in the electronic structure of $SrBi_2Ta_2O_9$ (SBT) and $SrBi_2Nb_2O_9$ (SBN) give higher spontaneous polarization (P_s) and ferroelectric transition temperatures.

1.3.3. *Dielectric losses and conductivity of APs*

Dielectric losses for APs increase significantly at temperatures above 400°C. This is usually associated with a high concentration of charge carriers (vacancies with a positive or negative charge) at high temperatures. Numerous examples show that doping of APs reduces dielectric losses, especially at high temperatures. The presence of impurities will affect the excess and hole conductivities differently at low temperatures. So, if the conductivity of a crystal is caused by the movement of ions in interstitial sites, then the introduction of impurity ions should increase the electrical conductivity especially strongly, since in this case there are few vacant sites in the lattice and impurity ions fall into the interstitial site (case 1). This provides both an increase in the number of carriers and a distortion of the lattice, which leads to low activation energies of impurities. On the contrary, in the case of hole conductivity, a small amount of impurities should not greatly increase the electrical conductivity (case 2). Impurities, in this case, can fill existing vacancies; this practically does not distort the lattice, and the conductivity may even decrease due to a decrease in the number of vacancies. It is also possible that impurities move in defective regions of the crystal, which corresponds to the case 1 considered above of excess conductivity. In general, the role of impurities in electrical conductivity in a number of cases is not so simple, and one can only say that the electrical conductivity of crystals is ultimately determined by the concentration of defects, the mobility of holes, the concentration of ions in interstitials and impurity ions.

It was shown in Ref. [469] that the dc conductivity in the temperature range of $\sim300° - \sim700°$C decreased from $\sim4.84 \times 10^{-5}$ S/cm (SBN) down to 2.42×10^{-6} S/cm (CaSBN, at $x = 0.075$) and to 2.09×10^{-6} S/cm (BaSBN, at $x = 0.05$), when the degree of doping the initial composition in A-position increased. At the same time, decreasing in B-position occurred to 8.56×10^{-6} S/cm for SBVN ($x = 0.05$) at a temperature of 500°C. The values of activation energy were in the range of 1.05–1.37 eV. From the values of the activation energy, it can be argued that the conductivity was due to the migration of oxygen vacancies [107, 456, 496]. The decrease in the values of activation energy with an increase in the degree of vanadium doping by 0.08 eV was caused by a decrease in the V–O binding energy (~627 kJ/mol), compared to the Nb–O binding energy (~703 kJ/mol). The introduction of Ba-ions leads to an increase in activation energy. The Ba–O binding energy (~562 kJ/mol), comparable to the Sr–O binding energy (~426 kJ/mol), could be the reason for this increase.

AP $Bi_4Ti_3O_{12}$ is a good ferroelectric with a relatively high Curie temperature. Nevertheless, it has a high anisotropy and electrical conductivity, which makes it difficult to polarize [108]. Doping it with donors such as W^{6+}, which replaces Ti^{4+}, leads to a strong increase in electric resistance, which facilitates polarization conditions [436]. AP $Bi_5FeTi_3O_{15}$ exhibits approximately the same behavior [400]. On the other hand, $ABi_4Ti_4O_{15}$ ($A = Ba, Sr, Ca$) also shows high electric resistance, which facilitates the polarization process [283].

The electrical conductivity of $Bi_4Ti_3O_{12}$ could be explained by the presence of a small polaron, activated by the Bi^{3+} cation, similar to the model that was applied to explain the conductivity of PZT [334]. In the result of a detailed analysis of current carriers, it was assumed that the Bi^{3+} cation had a valence state greater than 3+ [462]. As a consequence, the change in the valence between Bi^{3+} and Bi^{4+} inside the $(Bi_2O_2)^{2+}$ layer or inside the perovskite layer explains the conductivity behavior [361]. This model also explained the directionality of conduction [406]. It was assumed that the direction of electrical conductivity was realized through the perovskite layers, and not through $(Bi_2O_2)^{2+}$, and represented very low conductivity of the perovskite layers, intercalated between conductive Bi-containing perovskite layers. Thus, combining low-conductivity perovskite layers with compounds having high-conductivity $BiTiO_3$ layers can significantly reduce the overall conductivity of the solid solution.

1.3.4. *Nonlinear electrical properties of APs*

The ceramic $SrBi_2Nb_2O_9$ is one of the most promising compounds for creating ferroelectric memory elements, since it is very stable under external influences [76]. A number of works have been devoted to the effect of substitution of ions in perovskite cells both in the crystallographic A-position and in B-position [106, 118, 154, 158, 295, 296, 335, 360].

Figure 1.13 shows the results of measuring the relative permittivity of $Sr_{1+x}Bi_{4-x}Ti_{4-x}Ta_xO_{15}$ ($x = 0$–1) ceramics at a frequency of 100 kHz [491]. The Curie temperature T_C of $SrBi_4Ti_4O_{15}$ (540°C) coincides with the literature data (550°C [165] and (530°C [399]). The temperature dependence of the permittivity (ε) of this ceramic has a very sharp maximum with the largest value of ε, associated with the Curie point T_C. By increasing strontium concentration, the

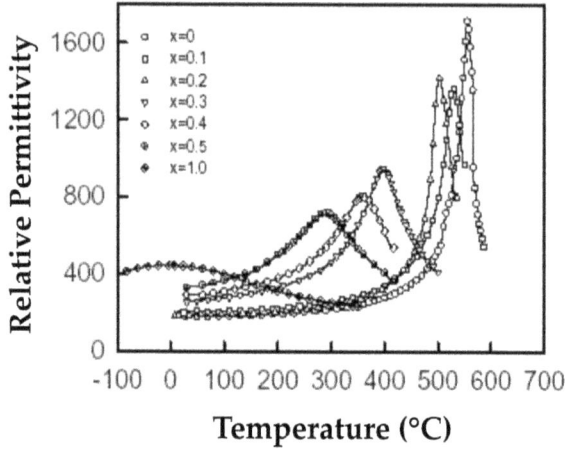

Fig. 1.13. Temperature dependences of the relative permittivity of ceramics $Sr_{1+x}Bi_{4-x}Ti_{4-x}Ta_xO_{15}$ ($x = 0-1$) at a frequency of $100\,kHz$.

peak of the dielectric permittivity shifts toward lower temperatures, the peaks expand and the value of ε decreases, which indicates an increase in diffuseness of phase transition.

Figure 1.14 shows the polarization versus electric field $(P-E)$ hysteresis loops of $Sr_{1+x}Bi_{4-x}Ti_{4-x}Ta_xO_{15}$ ($x = 0-1$) [491]. At an electric field strength of $100\,kV/cm$, the hysteresis loops for all samples are typical of conventional ferroelectrics at room temperature. It is important to note that in samples with a low content of strontium, such as $SrBi_4Ti_4O_{15}$ ($x = 0$) or $Sr_{1.2}Bi_{3.8}Ti_{3.8}Ta_{0.2}O_{15}$ ($x = 0.2$), the hysteresis loops do not have saturation regions and a much larger field strength must be applied to achieve saturation of the electric field. Materials with a concentration of $x = 0.4-0.5$ are saturated at an electric field strength of $100\,kV/cm$; their residual polarization and coercive field lie in the ranges of 7–$8\,\mu C/cm^2$ and 37–$47\,kV/cm$, respectively. Therefore, the substitution of Sr^{2+} and Ta^{5+} by Bi^{3+} and Ti^{4+} ions, respectively, can reduce the coercive field for $Sr_{1+x}Bi_{4-x}Ti_{4-x}Ta_xO_{15}$ ($x = 0.4-0.5$) compounds, which have a sufficiently large residual polarization and a relatively small coercive field, which are very important for ferroelectric applications. Thus, for $SrBi_4Ti_4O_{15}$ ceramics, the substitution of ions in A- and

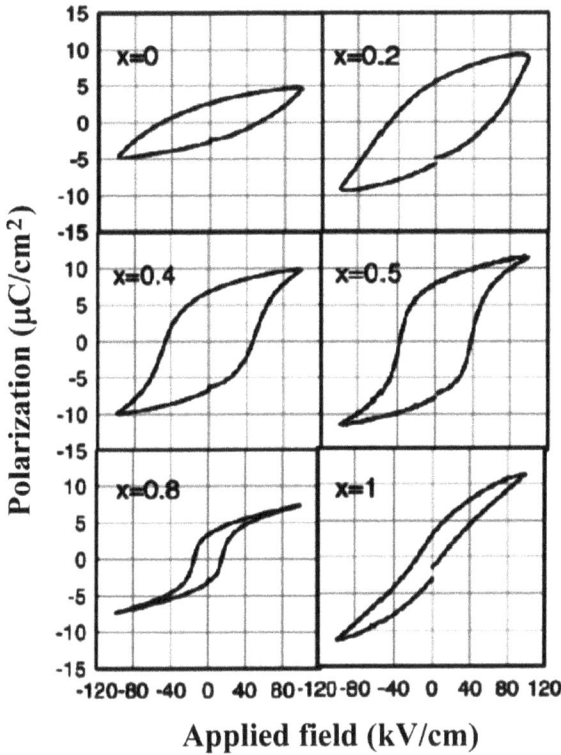

Fig. 1.14. Hysteresis loops for ceramics $Sr_{1+x}Bi_{4-x}Ti_{4-x}Ta_xO_{15}$ ($x = 0 - 1$).

B-positions with other ions makes it possible to improve the properties of bismuth-containing layered structures.

Figure 1.14 shows the hysteresis loops for $SrBi_4Ti_4O_{15}$ and $Sr_{1.1}Bi_{3.9}Ti_{3.9}Ta_{0.1}O_{15}$. The hysteresis loop for $SrBi_4Ti_4O_{15}$ is symmetrical up to a field strength of $120\,kV/cm$. For $Sr_{1.1}Bi_{3.9}Ti_{3.9}Ta_{0.1}O_{15}$ ceramics, when a $P - E$ field is applied, the loop shifts noticeably along the E-axis, demonstrating the presence of a relatively large internal field [104]. Many modified ferroelectric materials show the existence of an internal electric field, the origin of which is attributed to the presence of dipolar defects [16, 53, 193]. The authors of Ref. [491] concluded that, in the case of $Sr_{1.1}Bi_{3.9}Ti_{3.9}Ta_{0.1}O_{15}$, dipolar defects can exist, since in this compound a small Sr^{2+} ion is replaced by a Ta^{5+} ion. However, for samples, containing Ta^{5+} $(x = 0.4, 0.5, 0.8, 1.0)$, this dipolar defect can be suppressed and no internal field observed.

1.3.5. *Magnetic properties of APs*

In many known APs, along with ferroelectric properties, magnetic properties are manifested. These materials are ferromagnetic-multiferroic materials capable of acquiring electric polarization under the action of an external magnetic field and a magnetic moment at the action of an external electric field. First, ferroelectric materials with a layered AP structure $Bi_5Ti_3FeO_{15}$, $Bi_6Ti_3Fe_2O_{18}$, $Bi_9Ti_3Fe_5O_{27}$ were studied in Ref. [169]. $Bi_5FeTi_3O_{15}$ (BFTO) is one of the most promising and therefore intensively studied multiferroic with an AP structure [25, 86, 179, 236, 268]. BFTO is a ferroelectric with a Curie temperature T_C of about 1020 K [268], while its magnetic characteristics define it as a superparamagnetic with a predominance of antiferromagnetic coupling. Weak ferromagnetism most likely comes from Fe^{3+}/Fe^{2+} ions.

A nonlinear magnetoelectric effect is also observed in BFTO [86]. Crystallographic studies of BFTO [59, 209, 267, 392] have shown that Fe^{3+} and Ti^{4+} ions are randomly distributed in the available octahedra. In [177–179, 391, 402], rare-earth elements have been used for the $Bi_5AFe_2Ti_3O_{18}$ system, where $A = La, Sm, Gd, Dy$. These compounds were of interest in connection with the retention of their ferroelectric and magnetic properties at high temperatures. It was shown that the Néel temperature in antiferromagnetic APs increases with an increase in the number of perovskite-like layers m from 80 K for $Bi_5Ti_3FeO_{15}$ to 300 K for $Bi_9Ti_3Fe_5O_{27}$. In contrast, $Bi_7Ti_3Fe_3O_{21}$ and $Bi_8Ti_3Fe_4O_{24}$, being antiferromagnets at low temperatures, show weak ferromagnetism at room temperature [390]. Studies of hyperfine interactions in APs have shown that at room temperature all APs with $m = 4 - 8$ are paramagnetic materials [180, 273]. On the whole, it should be concluded that the study of the magnetic properties of APs is only the initial stage of their development, and a greater accumulation of experimental data is required to obtain definite conclusions.

1.4. Compounds with Integer m

Reports on the synthesis of several APs with different values of m, such as $Bi_7CaNaNb_6O_{27}$, $Bi_9Ti_4NbWO_{27}$, $Bi_9FeTiNb_4O_{27}$, $Bi_{11}CaTi_8NbO_{36}$, $Bi_{12}FeTi_7NbO_{36}$, $Bi_{13}Ca_2GaTi_{11}O_{45}$, $Bi_9PbCr Ti_7O_{30}$, $Bi_{14}ThFe_4Ti_8O_{45}$, were published in Refs. [116, 117, 206, 437, 438]. For all these structures, X-ray diffraction data were obtained and, on their basis, the cell parameters were determined. The formulae of the compounds obtained, as well as the structural parameters, are shown in Table 1.2. Interestingly, almost all of the compounds obtained contain four different cations, which means that in these compounds one or more atomic positions are shared between atoms of different types. Another new structure, $Bi_5AgNb_4O_{18}$ with $m = 2$, was found in the process of studying the phase equilibrium in the Bi_2O_3–Ag_2O–Nb_2O_5 system [464]. In the result of the refinement of the Rietveld structure on the basis of neutron diffraction data, it was found that the compound is isostructural with Bi_3TiNbO_9, $Bi_5NaNb_4O_{18}$ and $Bi_5KNb_4O_{18}$. The structure was refined in space group $A2_1am$, and the cell parameters are also present in the Table 1.2. By using $Bi_2Sr_2Nb_2TiO_{12}$ as an example, the authors of Ref. [264] demonstrated the possibility of isovalent and non-isovalent substitution in the A- and B-positions of perovskite layers, leading to the formation of new APs.

In particular, a double non-isovalent substitution of Sr by Na at the A-position and Ti by Ta at the B-position results in a new compound $Bi_2SrNaNb_2TaO_{12}$, and through a single non-isovalent substitution at the B-position, new compositions form such as $Bi_2Sr_2Nb_{2.5}Fe_{0.5}O_{12}$ and $Bi_2Sr_2Na_{2.6}Zn_{0.33}O_{12}$.

In the process of searching for new ferromagnetic materials, several new Aurivillius phases were synthesized, including Mn- and Fe-ions, such as $SrBi_3Nb_2FeO_{12}$, $Bi_2Sr_2Nb_2MnO_{12-\delta}$, $Bi_2Sr_{1.4}La_{0.6}Nb_2MnO_{12-\delta}$, as well as the epitaxial films with $m = 6$ and composition $Bi_7Mn_{3.73}Ti_{2.25}O_{21-\delta}$ [274, 388, 389, 479, 514]. The presence of ferroelectric and ferromagnetic properties at temperatures below room temperature was confirmed for the $SrBi_3Nb_2FeO_{12}$ composition obtained by combining $SrBi_2Nb_2O_9$ and

Table 1.2. New Aurivillius phases, their compositions, space groups and derived unit cell parameters.

Compound	m	Space group	Cell parameters, Å			Refs.
			a	b	c	
$Bi_7CaNaNb_6O_{27}$	2	$A2_1am$	5.4517(4)	5.4877(1)	24.8963(1)	[441]
$Bi_9Ti_4NbWO_{27}$	2	$A2_1am$	5.3887(1)	5.4078(6)	24.9474(9)	[441]
$Bi_9FeTiNb_4O_{27}$	2	$A2_1am$	5.4197(1)	5.4463(9)	25.2498(6)	[441]
$Bi_{11}CaTi_8NbO_{36}$	3	$B2cb$	5.4040(1)	5.4287(1)	32.8058(2)	[441]
$Bi_{12}FeTi_7NbO_{36}$	3	$B2cb$	5.4162(5)	5.4496(3)	32.9204(1)	[441]
$Bi_{13}Ca_2GaTi_{11}O_{45}$	4	$A2_1am$	5.4094(1)	5.4277(1)	40.9950(1)	[441]
$Bi_9PbCrTi_7O_{30}$	4	$A2_1am$	5.4240(2)	5.4320(1)	41.0437(1)	[441]
$Bi_{14}ThFe_4Ti_8O_{45}$	4	$A2_1am$	5.4300(2)	5.4451(7)	41.2800(6)	[441]
$Bi_{2.25}Ca_{0.5}$ $Na_{0.25}Nb_2O_9$	2	$A2_1am$	5.4845	5.4549	24.9195	[173]
$SrBi_3Ti_2NbO_{12}$	3	$I4/mmm$	3.850	—	33.21	[336]
$PbBi_3Ti_2NbO_{12}$	3	$I4/mmm$	3.865	—	33.52	[336]
$Bi_2Sr_2Nb_2ZrO_{12}$	2	$A2_1am$	5.4845	5.4549	24.9195	[105]
$Bi_{2.25}Ca_{0.5}Na_{1.25}$ Nb_3O_{12}	3	$Fmmm$	5.4574(7)	5.4844(3)	32.711(6)	[173]
$Bi_2CaNaNb_3O_{12}$	3	$Fmmm$	5.4474(7)	5.4770(3)	32.722(6)	[173]
$Bi_2CaNa_2Nb_4O_{12}$	4	$A2_1am$	5.4584	5.4833	40.534	[173]
$Bi_3Fe_{0.5}Nb_{1.5}O_9$	2	$Fmmm$	5.4338(9)	5.4626(8)	25.389(1)	[513]
$Gd_{0.5}Nd_{0.5}$ $Bi_3Ti_3O_{12}$	3	$I4/mmm$	3.816(2)	—	32.82(6)	[336]
$Ca_{0.25}Th_{0.25}$ $Bi_3Ti_3O_{12}$	3	$I4/mmm$	3.826(4)	—	32.81(5)	[336]
$Th_{0.25}Bi_{3.5}Ti_{2.5}$ $Ga_{0.5}O_{12}$	3	$I4/mmm$	3.823(4)	—	32.95(8)	[336]
$Bi_5AgNb_4O_{18}$	2	$A2_1am$	5.4915(2)	5.4752(2)	24.9282(8)	[128]
$Bi_{2.54}Li_{0.38}Nb_2O_9$	2	$Cmc2_1$	24.840(5)	5.449(6)	5.450(9)	[41]
$Bi_2SrNaNb_2TaO_{12}$	3	$I4/mmm$	3.896(1)	—	32.839(7)	[105]
$Bi_2Sr_2Nb_{2.5}$ $Fe_{0.5}O_{12}$	3	$I4/mmm$	3.910(1)	—	32.245(5)	[105]
$Bi_2Sr_2Nb_{2.67}$ $Zn_{0.33}O_{12}$	3	$Fmmm$	5.528(1)	5.535(1)	33.515(9)	[105]
$SrBi_3Nb_2FeO_{12}$	3	$Fmmm$	5.488	5.416	29.62	[174]
$Bi_2Sr_2Nb_2MnO_{12-\delta}$	3	$Fmmm$	5.5243(4)	5.5346(3)	32.9996(9)	[250]
$Bi_2W_2O_9$	2	$Pna2_1$	5.440(1)	5.413(1)	23.740(5)	[37]

$BiFeO_3$ compounds in a ratio of 1:1 [388, 389]. The piezoelectric properties of the compound were studied by the resonance method. In the result of magnetic measurements at room temperature, a narrow hysteresis loop was obtained, indicating antiferromagnetic ordering. A significant magnetoelectric response was found at 77 K. A study [479] of the properties of $Bi_2Sr_2Nb_2MnO_{12-\delta}$ showed that this is a paramagnetic compound, which also has semiconductor properties. Based on the diffraction data, it was established that the Mn and Nb atoms in the structure were completely ordered. Thin films of $Bi_7Mn_{3.73}Ti_{2.25}O_{21}$ [514] exhibit ferromagnetic properties at temperatures below 55 K; however, no ferroelectric properties were found in them.

Another interesting new structure from the APs series with $m = 2$ was described in Ref. [55]. The compound with the formula $Bi_2W_2O_9$ belongs to the cation-deficient APs, in which Bi_2O_2 layers alternate with ReO_3-like layers. The structure of this compound was determined based on X-ray diffraction data, obtained from single crystal samples. The compound had a rhombic $Pna2_1$ cell with the parameters: $a = 5.440(1)$ Å, $b = 5.413(1)$ Å, $c = 23.740(5)$ Å. The W_2O_7 layers consisted of two layers of octahedra, and the atoms W were displaced in each layer from the center along a-axis, and these displacements were antiparallel in two layers of the octahedra. In ReO_3-like layers, the structure was distorted strongly, and these distortions were associated with the antiparallel displacements of the atoms W.

For the first time, APs containing lithium atoms were obtained, and Li can both enter into the composition of these compounds and form intercalated compounds. In particular, several new APs were synthesized with the general formula $Bi_{2.5+x}Li_{0.5-3x}Nb_2O_9$ $(0.04 < x < 0.08)$ [269]. X-ray diffraction analysis showed that the structure of the compound with $x = 0.04$ is orthorhombic $Cmc2_1$, the parameters are given in Table 1.2.

In Ref. [68], the preparation of AP $Li_2Bi_4Ti_3O_{12}$ intercalated with lithium was reported. An X-ray diffraction study showed that Li atoms occupy positions in the gap between the perovskite and fluorite-like layers, and in the result of studying the local structure of the compound, it was found that intercalation leads to an increase in distortions of the local environment of the Ti atom.

A series of Bi_2WO_6-derived APs was obtained by replacing Bi atoms with lanthanide atoms [33], which had the general formula $Bi_{2-x}Ln_xWO_6$, where $0.3 < x < 1.3$. The cell parameters of one of the compounds of this series $Bi_{0.7}Yb_{1.3}WO_6$, obtained on the basis of neutron diffraction, are equal to: $a = 8.1070\,\text{Å}$, $b = 3.7048\,\text{Å}$, $c = 15.8379\,\text{Å}$, $\beta = 103.548°$.

1.5. Compounds with Non-integer m

A detailed analysis of the structural distortions of APs with $m = 1.5$ was carried out using the example of $Bi_5Ti_{1.5}W_{1.5}O_{15}$. The structure of the compound $Bi_5TiNbWO_{15}$ was studied in detail by X-ray and neutron diffraction [370] and refined by the Rietveld method. At room temperature, the structure was defined as rhombic $I2cm$ with the following unit cell parameters: $a = 5.4231\,\text{Å}$, $b = 5.4027\,\text{Å}$, $c = 41.744\,\text{Å}$. The parameters a and b were close to those, observed in compounds of the APs with $m = 1$ and $m = 2$. The arrangement of atoms in the cell is shown in Figure 1.15. Structural distortions are typical of APs and are characterized by rotation and distortion of oxygen octahedra. Based on X-ray diffraction data, it was revealed that the cations were completely ordered in a single perovskite layer, while disorder of Ti and Nb atoms was observed in a double perovskite layer [413]. Based on the complete determination of the structure, carried out by the Rietveld method, three types of structural distortions characteristic of this structural type were analyzed, namely: (i) rotation of oxygen octahedra around the a-axis; (ii) rotation of oxygen octahedra around the b-axis; (iii) displacement of atoms along the polar c-axis. It was found that the distinguishing feature of the $Bi_5Ti_{1.5}W_{1.5}O_{15}$ structure was the disappearance of octahedral rotations in single perovskite layers while they remained in double layers, as shown in Figure 1.16. Moreover, the changes of distances were observed between cations, located at the centers of oxygen octahedra, and their oxygen environment.

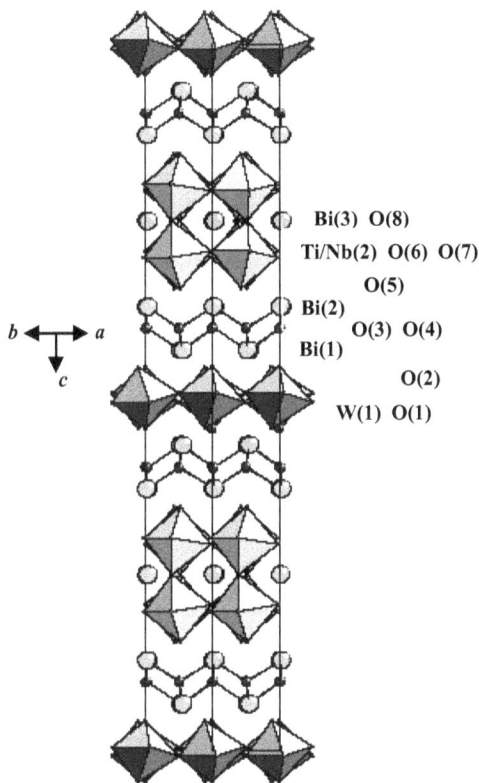

Fig. 1.15. Crystal structure of $Bi_5TiNbWO_{15}$: View along direction [110].

Recently, the previously small class of compounds with non-integer m, especially with $m = 1.5$, has been significantly expanded. For the first time, layered perovskite-like oxides $Bi_5TiNbWO_{15}$ and $Na_{0.5}Bi_{4.5}Nb_2WO_{15}$, which belong to Aurivillius phases with non-integer $m = 1.5$, were described in [200]. In the $Bi_5TiNbWO_{15}$ structure, layers of the perovskite-like Bi_2WO_6 structure of one oxygen octahedron thickness alternate with layers of two oxygen octahedrons thickness of another known composition, Bi_3TiNbO_9. The $Bi_5Ti_{0.5}W_{0.5}Nb_2O_{15}$ structure is obtained from the $Bi_5TiNbWO_{15}$ structure by reducing the number of Ti and W atoms and increasing the number of Nb atoms. The $Bi_5Nb_3O_{15}$ composition can be obtained due to complete replacement of Ti and W by Nb atoms.

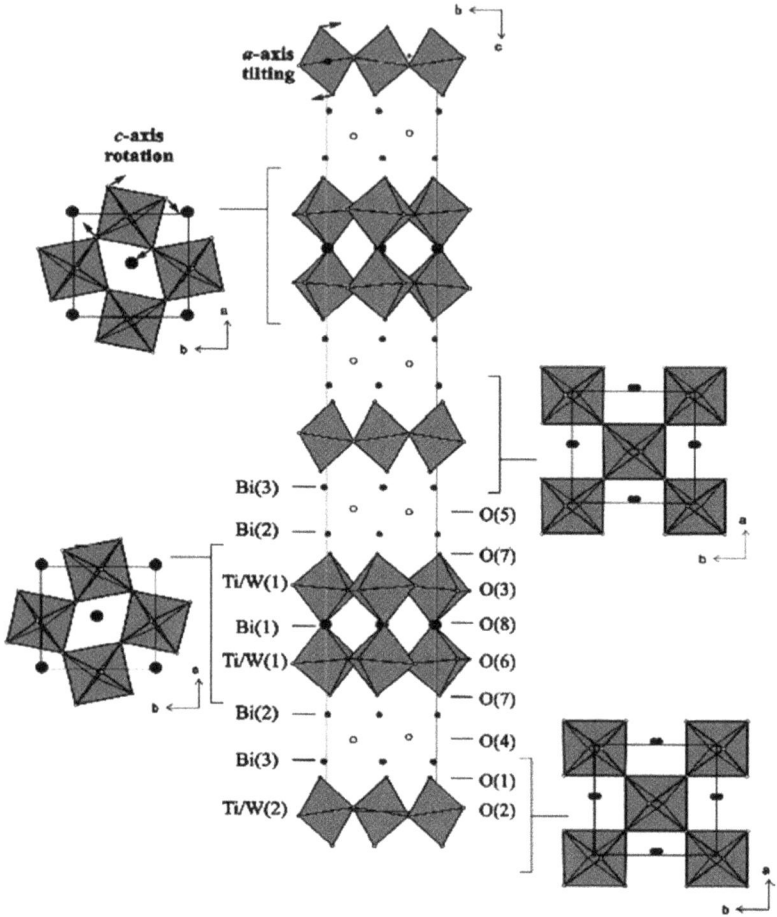

Fig. 1.16. Crystal structure of $Bi_5TiNbWO_{15}$ in plane.

X-ray diffraction analysis of all obtained compounds was carried out. X-ray patterns were indexed in a tetragonal cell and the obtained parameters are given in Table 1.3. It should be noted, however, that X-ray diffraction patterns of $Bi_5TiNbWO_{15}$ show line broadening, which indicates a rhombic distortion of the structure.

The dielectric characteristics of all three compounds, $Bi_5TiNbWO_{15}$, $Bi_5Ti_{0.5}W_{0.5}Nb_2O_{15}$ and $Bi_5Nb_3O_{15}$, were studied in Refs. [105, 243]. The behavior of the relative permittivity and dielectric loss curves showed that all of these compounds are ferroelectric,

Table 1.3. Aurivillius phases with non-integer m, their compositions, space groups and unit cell parameters.

Compound	m	Space group	Cell parameters, Å		
			a	b	c
$Bi_5TiNbWO_{15}$	1.5	$I4/mmm$	3.832	—	20.97
$Bi_5Ti_{0.5}W_{0.5}Nb_2O_{15}$	1.5	$I4/mmm$	3.853	—	21.020
$Bi_5Nb_3O_{15}$	1.5	$I4/mmm$	3.85	—	20.90
$Bi_4LaNb_3O_{15}$	1.5	$Cmm2$	5.469(4)	5.471(2)	20.864(9)
$Bi_4PrNb_3O_{15}$	1.5	$Cmm2$	5.464(1)	5.466(1)	20.865(4)
$Bi_4NdNb_3O_{15}$	1.5	$Cmm2$	5.461(1)	5.462(1)	20.841(4)
$Bi_4LaTa_3O_{15}$	1.5	$Cmm2$	5.476(3)	5.478(2)	20.820(6)
$Bi_{14}PbNb_8WO_{45}$	1.5	$Cmm2$	5.4482(3)	5.4563(2)	20.8380(1)
$Bi_{10}Ti_{0.5}Nb_5W_{0.5}O_{30}$	1.5	$Cmm2$	5.4505(1)	5.4541(2)	20.9381(2)
$Bi_4Pb_{1.5}Ti_{4.5}O_{16.5}$	4.5	$Cmm2$	5.4434(1)	5.4506(1)	45.527(2)
$Bi_5Ca_{0.5}GaTi_{3.5}O_{16.5}$	4.5	$Cmm2$	5.3868(7)	5.4185(8)	45.267(5)

but none of these compounds could obtain hysteresis loops, which could be due to relatively high dielectric losses.

A series of new compounds has recently been synthesized from a series of Aurivillius phases with $m = 1.5$, containing lanthanides with the formulae $Bi_4LnNb_3O_{15}$ (where Ln = La, Pr, Nd) and $Bi_4LaTa_3O_{15}$ [263]. The study of these compounds by X-ray diffraction and electron microscopy showed that they are rhombic, and the parameters of their cells are given in Table 1.3.

All structures consist of AP fragments with $m = 1(Bi_2NbO_6)$ and $m = 2(Bi_2LnNb_2O_9)$, alternating in order along c-axis. In all compounds, it was not possible to detect second harmonic generation on laser radiation with a wavelength of 1064 nm, and according to the results of dielectric measurements, no ferroelectric–paraelectric phase transition was observed in the temperature range from 30 to 900°C, which may be evidence of a centrosymmetric structure. It should be noted that in these compounds, the value of the cell parameter c, determined on the basis of X-ray diffraction data, turns out to be two times less than the similar parameters of some other representatives of APs with $m = 1.5$, for example, $Bi_5TiNbWO_{15}$ and $Bi_5Ti_{1.5}W_{1.5}O_{15}$. Owing to this circumstance, the authors performed additional studies of materials by electron diffraction and found reflections, corresponding to a double cell. The impossibility of detecting these reflections by the X-ray diffraction method was explained by the smallness of the structural distortions, reduced mainly to rotations of oxygen octahedra, which gave a small contribution to the X-ray diffraction.

The manufacture of two new compounds with $m = 1.5$ and formulae $Bi_{14}PbNb_8WO_{45}$, $Bi_{10}Ti_{0.5}Nb_5W_{0.5}O_{30}$ was reported in Ref. [438], and two new compounds with $m = 4.5$ were described in Ref. [362]. All structures are rhombic $Cmm2$, and their cell parameters are given in Table 1.3.

1.6. Distortions of Ideal Structure

In ferroelectric compounds of the perovskite family, as well as layered materials such as APs, the properties are closely related to small distortions of an ideal, highly symmetrical structure. In this regard, obtaining and analyzing information about such distortions can provide a key to understanding the nature of ferroelectricity in these materials and developing ways to obtain new materials with improved properties. The development of methods for structural studies, and the improvement of equipment, which leads to a significant increase in the accuracy of determining structural characteristics, opens up new possibilities in the study of fine features of the structure of materials. Therefore, interest has recently increased in studying and refining the structure of new and known APs.

The complete refinement of the $Bi_{0.7}Yb_{1.3}WO_6$ structure was carried out in space group $A2$ by the Rietveld method [33]. In the result of this research, a new unusual type of atomic ordering was discovered. Unlike Bi_2WO_6, where the WO_4 layer consists of a two-dimensional sequence of vertex-linked octahedra, the octahedra in $Bi_{0.7}Yb_{1.3}WO_6$ are coupled both along vertices and edges, as shown in Figure 1.17. Moreover, some ordering of the arrangement of Bi and Yb was observed, in which the Yb atoms predominantly occupy positions near oxygen atoms, forming edges common for two neighboring oxygen octahedra. This combination of cationic

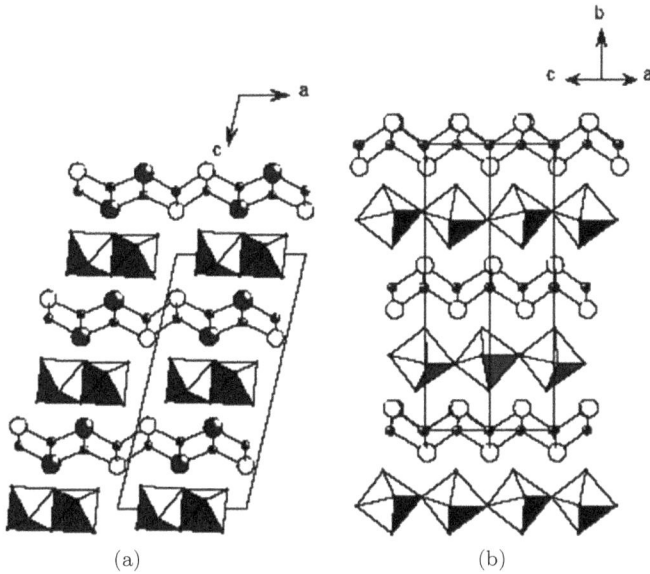

Fig. 1.17. (a) $Bi_{0.7}Yb_{1.3}WO_6$ structure with alternating $(Bi, Yb)_2O_2$ layers and WO_4 layers with edge-linked octahedra, (b) Bi_2WO_6 structure with WO_4 layers, consisting of vertex-sharing octahedra.

ordering, combined with the distortions, associated with the unusual packing of the octahedra, leads to a reduction in symmetry from $A2/m$ to the polar group $A2$. These factors, as well as short W–W distances, causing a repulsive interaction between them, contribute to significant distortions of the WO_6 octahedra themselves.

Over the past decades, the structures of the compounds with $m = 2$ and $m = 4$ have been very intensively studied in order to reveal the regularities of their structures. Almost all compounds of these series, described in the literature at room temperature, have an orthorhombic cell with the space group $A2_1am$ and close parameters, which are given in Table 1.4. The exception is $BaBi_2Ta_2O_9$, which was refined in the highly symmetric space group $I4/mmm$, and its cell parameters are $a = 3.92650(8)$ Å and $c = 25.5866(8)$ Å [359].

It has been noted that most of the APs with Ba atoms in the A-position of the perovskite layer, in contrast to their analogs, in which Sr, Ca and Pb atoms are in the A-position, exhibit relaxor

Table 1.4. Cell parameters of APs with $m = 2$ and space group $A2_1am$.

Compound	a, Å	b, Å	c, Å
$CaBi_2Ta_2O_9$	5.4625(1)	5.4286(1)	24.9450(6)
$SrBi_2Ta_2O9$	5.5224(2)	5.5241(2)	25.0264(5)
$Bi_{2.5}Na_{0.5}Nb_2O_9$	5.4937(3)	5.4571(4)	24.9169(14)
$Bi_{2.5}K_{0.5}Nb_2O_9$	5.5005(8)	5.4958(8)	25.2524(16)
$Bi_{2.5}Na_{0.5}Ta_2O_9$	5.4763(4)	5.4478(4)	24.9710(15)
$Bi_2AgNb_4O_{18}$	5.4915(2)	5.4752(2)	24.9282(8)
$Bi_2SrNb_2O_9$	5.5193(3)	5.5148(3)	25.0857(6)
$Bi_2CaNb_2O_9$	5.4833(1)	5.4423(1)	24.8984(6)
$PbBi_2Nb_2O_9$	5.4909(1)	5.4998(1)	25.5313(2)
$Bi_3Ti_{1.5}W_{0.5}O_9$	5.40181(2)	5.37274(1)	24.9388(1)
$Bi_{2.25}Ca_{0.5}Na_{0.25}Nb_2O_9$	5.4845	5.4549	24.9195
$K_{0.5}La_{0.5}Bi_2Ta_2O_9$	5.512(2)	5.504(2)	25.072(4)

properties. Recently, in order to elucidate the structural reasons for such differences, a series of comparative studies were carried out in series of APs with different atoms in the A-position. In the course of these studies, some structural features were revealed that are more or less characteristic of all Aurivillius phases.

Thus, the atoms, located in the A-positions and B-positions are displaced along the c-axis in the direction toward the oxygen atoms, lying in the fluorite-like layers. As one moves away from the fluorite-like layers, this shift becomes smaller. This expansion of the cationic sublattice is combined with contraction of the anionic sublattice, leading to the formation of short B–O bonds. The shift of cations in A-position also occurs along the polar a-axis.

On the basis of powder diffraction patterns, the structures of three compounds of the $ABi_2Ta_2O_9$ series, where A = Ca, Sr, Ba, were refined (see Figure 1.18). The study made it possible to trace the influence of the value of A-cation on the degree of structural distortions. The figure clearly shows how the distortion of TaO_6 octahedra increases as the size of the cation in the A-position decreases. A similar comparative study for $Bi_{2.5}Na_{0.5}Nb_2O_9$ and $Bi_{2.5}K_{0.5}Nb_2O_9$ was described in Ref. [41]. Later, two other compounds of this series were investigated, namely, $Bi_{2.5}Na_{0.5}Ta_2O_9$ [42] and $Bi_{2.5}Ag_{0.5}Nb_2O_9$ [464]. Here, the same regularity was

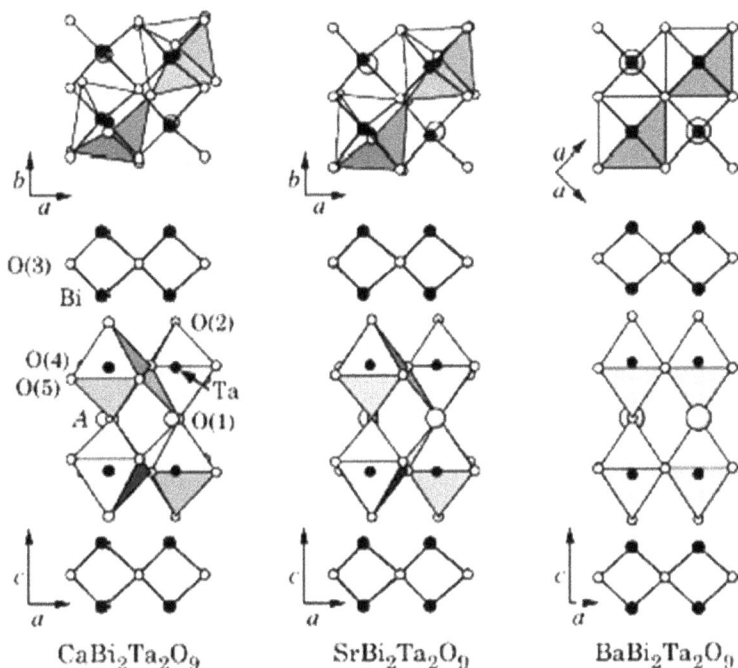

Fig. 1.18. Crystal structure of $ABi_2Ta_2O_9$, where A = Ca, Sr, Ba.

observed as in the previous case and rhombic distortions increased in the sequence: A = K−Ag−Na.

As regards the rotation of octahedra, there is no obvious dependence on the size of the A-cation, and the rotation of octahedra in $Bi_{2.5}Ag_{0.5}Nb_2O_9$ along the c-axis turns out to be greater than in $Bi_{2.5}Na_{0.5}Nb_2O_9$ and $Bi_{2.5}K_{0.5}Nb_2O_9$, which may be due to differences in the electronic structure of these atoms. The displacement of Nb atoms along the c-axis in respect to the centers of oxygen octahedra is the largest for the K-containing compound and decreases by increasing A-ion radius. Compared to the isostructural Nb analog, the structure of $Bi_{2.5}Na_{0.5}Ta_2O_9$ is less distorted, and calculations, using the valence sum method, show that the Ta − O bonds are somewhat stronger, compared to the Nb − O bonds.

Similar studies were carried out for a number of Aurivillius phases with m = 4, such as $BaBi_4Ti_4O_{15}$, $CaBi_4Ti_4O_{15}$, $SrBi_4Ti_4O_{15}$ and

Table 1.5. Cell parameters of APs with $m = 4$ (space group $A2_1am$) and $m = 5$ (space group $B2ab$).

Compound	a, Å	b, Å	c, Å	Refs.
$BaBi_4Ti_4O_{15}$	5.4433(4)	5.4319(6)	41.6941(3)	[485]
$CaBi_4Ti_4O_{15}$	5.4234(2)	5.4021(5)	40.5935(3)	[485]
$SrBi_4Ti_4O_{15}$	5.4510(5)	5.4415(5)	41.0233(13)	[90]
$PbBi_4Ti_4O_{15}$	5.4267(1)	5.4458(1)	41.4121(12)	[90]
$Bi_{2.5}Na_{2.5}Nb_4O_{15}$	5.5095(5)	5.4783(5)	40.553(3)	[413]
$Ca_2Bi_4Ti_5O_{18}$	5.4251(2)	5.4034(1)	48.486(1)	[118]
$Sr_2Bi_4Ti_5O_{18}$	5.4650(2)	5.4625(3)	48.852(1)	[118]
$Ba_2Bi_4Ti_5O_{18}$	5.4988(3)	5.4577(4)	49.643(1)	[118]
$Pb_2Bi_4Ti_5O_{18}$	5.4980(2)	5.4701(2)	50.352(1)	[118]

$PbBi_4Ti_4O_{15}$ [401, 414], and also with $m = 5$, such as $Ba_2Bi_4Ti_5O_{18}$, $Ca_2Bi_4Ti_5O_{18}$, $Sr_2Bi_4Ti_5O_{18}$ and $Pb_2Bi_4Ti_5O_{18}$ [171]. The parameters of several APs with $m = 4$ and $m = 5$, whose structure has been studied recently, are shown in Table 1.5.

A comparative study of the structures of $BaBi_4Ti_4O_{15}$ and $CaBi_4Ti_4O_{15}$, which represent two extreme cases with respect to the radius of the cation in the A-position, was carried out in Ref. [414]. In terms of layer packing along the c-axis, both structures are similar and exhibit features characteristic of all APs. The most obvious differences in the structures are due to the degree of their distortion, which is much higher for the Ca-containing compound. In $CaBi_4Ti_4O_{15}$, the oxygen octahedra are rotated around the a- and c-axes alternately in different directions, while these distortions are very small in $BaBi_4Ti_4O_{15}$, which makes it possible to refine this structure in more highly symmetric space groups. $BaBi_4Ti_4O_{15}$ also exhibits partial substitution of Bi atoms by Ba atoms in fluorite-like layers. The relaxor properties of the compound are explained in the work by the fact that the positions of the Bi and Ba atoms in the perovskite layers differ significantly, which leads to the formation of microregions with varying degrees of structural distortions.

The influence of the number of layers on structural distortions was studied by using the examples of a lot of compounds: $Bi_{2.5}Na_{0.5}Ta_2O_9$ ($m = 2$), $Bi_{2.5}Na_{1.5}Nb_3O_{12}$ ($m = 3$) and $Bi_{2.5}Na_{2.5}Nb_4O_{15}$ ($m = 4$) [42].

As the number of layers increases, the parameters a and b also increase, contributing to an increase in the stress in the structure. Moreover, electron microscopy data show that an increase in the number of layers leads to an increase in the number of defects in the crystal packing.

In the structure of another compound with $m = 4$ ($CaCd_2$ $Bi_2Nb_4O_{15}$), the presence of NbO_6 octahedra, coupled by common edges with vertex-coupled octahedra, was found.

1.7. Structural Disorder

In compounds belonging to the APs, most attention has been recently paid to the problems of structural disorder. A number of APs are characterized by such a type of structural disorder, in which atoms of two types, with different probabilities, occupy two different atomic positions, one of which is in the fluorite layer, and the other in the perovskite layer. Diffraction methods were used to study this type of structural disorder in compounds with the general formula $Bi_2ANb_2O_9$, where A = Ba, Sr, Ca and Pb [37, 170]. The distortions of the $Bi_2BaNb_2O_9$ structure are small, and it was successfully refined in the $I4/mmm$ space group. The cell parameters were $a = 3.93$ Å and $c = 25.6582$ Å. $Bi_2SrNb_2O_9$, $Bi_2CaNb_2O_9$ and $PbBi_2Nb_2O_9$ crystallize in the space group $A2_1am$, and their cell parameters are given in Table 1.6.

The combination of X-ray and neutron diffraction made it possible to detect in all these compounds a partially mixed distribution of A-type and Bi cations according to their positions. The degree of this

Table 1.6. Cell parameters of APs with $m = 3$ and cell $I4/mmm$.

Compound	a(Å)	c(Å)
$Bi_2La_2Ti_3O_{12}$	3.83166(4)	33.0139(5)
$Bi_2Pr_2Ti_3O_{12}$	3.80953(2)	32.8143(3)
$Bi_2Nd_2Ti_3O_{12}$	3.80620(2)	32.7647(3)
$Bi_2Sm_2Ti_3O_{12}$	3.79570(3)	32.7099(4)
$Bi_2Sr_{1.4}La_{0.6}Nb_2MnO_{12}$	3.89970(7)	32.8073(9)

disorder increases in the order Ca < Sr < Ba and is determined both by the size of the cations and the valence factor. It was also found that the degree of disorder in $PbBi_2Nb_2O_9$ is affected by the sample cooling rate. The disordering of cations was studied in a number of APs with $m = 3$, including various elements from the lanthanide series with the formulae $Bi_2Ln_2Ti_3O_{12}$, where $Ln =$ La, Pr, Nd and Sm [163], as well as $Bi_3LaTi_3O_{12}$[152]. All $Bi_2Ln_2Ti_3O_{12}$ compounds had a tetragonal structure $I4/mmm$. Their cell parameters, given in Table 1.6, and Figure 1.19, show a schematic representation of their structure. $Bi_3LaTi_3O_{12}$, whose structure was also initially determined as tetragonal, has a non-centrosymmetric space group $B2cm$ according to refined data, and a cell with the following parameters: $a = 5.450(1)$ Å, $b = 5.4059(6)$ Å, $c = 32.832(3)$ Å [69]. It was found that Bi atoms and lanthanide atoms are disordered over the corresponding atomic positions. For $Bi_3LaTi_3O_{12}$, the degree of disorder is estimated to be about 20% (that is, about 20% of the bismuth atoms occupy A-positions in the perovskite layer). However, when the size of the lanthanide atoms increases, the degree of disorder gradually decreases. A structural study of $Bi_2SrCaNb_2TiO_{12}$, $Bi_2Sr_{1.5}Ca_{0.5}Nb_2TiO_{12}$, $Bi_2Sr_2Nb_2TiO_{12}$ and

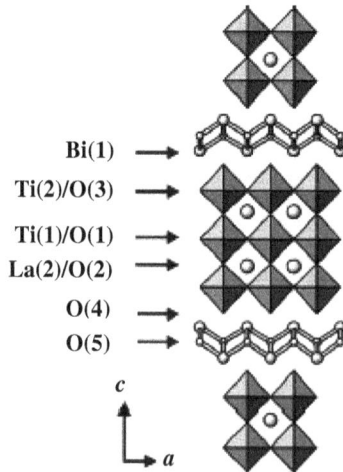

Fig. 1.19. Schematic representation of the structure of $Bi_2Ln_2Ti_3O_{12}$.

$Bi_2Sr_{1.5}Ba_{0.5}Nb_2TiO_{12}$, carried out by the method of combined analysis of the data of X-ray diffraction and neutron diffraction [145], revealed structural disorder in the form of mixed occupancies of Bi sites and *A*-type atomic sites in the perovskite layer. Disorder was also observed in the Nb and Ti positions.

In this case, an inverse relationship was also found between the degree of disorder and the radius of the cation of *A*-type. Ca atoms predominantly occupy *A*-type positions in the perovskite layer, while Ba atoms intensively settle on Bi positions, which helps to reduce the interlayer stress. The degree of disorder at the level of 20% is also observed in solid solutions $Bi_{2-x}Sr_{2+x}Ti_{1-x}Nb_{2+x}O_{12}$ (where $0 < x < 0.8$) [152] and $Bi_2Sr_{1.4}La_{0.6}Nb_2MnO_{12}$ [274]. These materials also have $I4/mmm$ tetragonal cells, the parameters of which are given in Table 1.6.

1.8. Ferroelectric Phase Transitions: Theory

For a formal description of ferroelectric transitions, as a rule, the Landau–Ginzburg–Devonshire phenomenological theory of phase transitions is used [82, 125, 126, 221, 222].

Within the framework of this theory, the thermodynamic parameters of the solid are considered for small deviations from the symmetrical state. In this case, values of some order parameter η are given, and the thermodynamic potential is represented as a function $\Phi(p, T, \eta)$. If the spontaneous polarization P is taken as the order parameter, then for the case of a uniaxial ferroelectric potential, we obtain

$$\Phi = \Phi_0 + \frac{\alpha}{2}p^2 + \frac{\beta}{4}p^4 - EP. \tag{1.2}$$

Based on the conditions of minimum thermodynamic potential: $\partial\Phi/\partial P = 0$ and $\partial^2\Phi/\partial P^2 > 0$, we get at $E = 0$:

$$\begin{cases} -\dfrac{\alpha}{\beta} = \dfrac{\alpha'(T - T_c)}{\beta}, & T < T_C, \\ 0, & T > T_C, \end{cases} \tag{1.3}$$

where

$$a(T) = \left(\frac{\partial \alpha}{\partial T}\right)_{TC} (T - T_{\mathrm{C}}) = \alpha'(T - T_{\mathrm{C}}); \quad \alpha' > 0, \quad \beta' > 0.$$

If in the thermodynamic potential, $E \neq 0$, then the solution of the equation, obtained at minimization of the potential, allows one to obtain the Curie–Weiss law for permittivity, as follows:

$$\varepsilon = \begin{cases} \dfrac{2\pi}{\alpha'_{T_c}(T - T_c)}, & T > T_{\mathrm{C}}, \\[3mm] -\dfrac{\pi}{\alpha'_{T_c}(T - T_c)}, & T < T_{\mathrm{C}}. \end{cases} \qquad (1.4)$$

In some ferroelectrics, such as BaTiO$_3$, phase transitions of the first order, close to the second order, occur. In this case, the expression for the thermodynamic potential is as follows:

$$\Phi = \Phi_0 + \frac{\alpha}{2}P^2 + \frac{\beta}{4}P^4 + \frac{\gamma}{6}P^6 - EP, \qquad (1.5)$$

where $\gamma(P, T) > 0$.

Minimization of potential (1.5) allows one to obtain expressions similar to (1.3) and (1.4), as follows:

$$P = \begin{cases} \dfrac{-\beta \pm \sqrt{\beta^2 - 2\alpha\gamma}}{\gamma}, & T < T_{\mathrm{C}}, \\[3mm] 0, & T > T_{\mathrm{C}}, \end{cases} \qquad (1.6)$$

$$\varepsilon = \begin{cases} \dfrac{2\pi}{\alpha(T)}, & T > T_{\mathrm{C}}, \\[3mm] \dfrac{-\pi}{2\alpha(T) - \beta P^2}, & T < T_{\mathrm{C}}. \end{cases} \qquad (1.7)$$

Ferroelectric transitions are also described using the dynamic theory [10, 126]. This theory assumes that the ferroelectric transition is due to the instability of one of the optical vibrational modes of the lattice. In the vicinity of the phase transition, the transverse mode of the optical vibration w_{r0} is described as

$$w_{r0}^2 = \begin{cases} \dfrac{\alpha}{\mu} = \dfrac{\alpha'_{T_c}(T - T_C)}{\mu}, & T > T_C, \\[3mm] \dfrac{4\alpha}{\mu} = \dfrac{4\alpha'_{T_c}(T - T_C)}{\mu}, & T < T_C. \end{cases} \quad (1.8)$$

At $T = T_C$, the value w_{r0}^2 turns to zero.

The Landau–Ginzburg–Devonshire theory has been continuously improved and modernized since its inception. The case of a one-dimensional crystal was generalized to multiaxial ferroelectrics. This was first performed in the 1950s by A. F. Devonshire for barium titanate [80,81,326]. In subsequent years, modifications of the theory were developed to describe phase transitions in antiferroelectric crystals [39]. The influence of free charge carriers on defects and dimensional defects was also considered [109, 230, 231].

The advantages of the thermodynamic theory include the simplicity of the mathematical technique, a wide range of applications and the possibility of establishing dependences between various macroscopic parameters of ferroelectrics. Its restrictions are caused by the macroscopic approach, which does not allow one to discuss the microscopic nature of the ferroelectric phase transition, as well as its applicability only to equilibrium processes.

1.9. Phase Transitions: Experiment

At present, there is great interest and active discussion connected with phase transitions between the low-temperature ferroelectric and high-temperature paraelectric phases in compounds, related to APs. In the compounds with $m = 2$, the ferroelectric phase usually has the space group $A2_1am$, and it passes into the $I4/mmm$ phase above the Curie temperature T_C. Recent diffraction studies using neutron and synchrotron sources have shown that these phase transitions can occur through the intermediate paraelectric orthorhombic phase *Amam* [151,254,255]. For APs with $m = 2$, the intermediate rhombic phase loses the mode, associated with the ferroelectric displacements of atoms, and also one of the modes associates with the rotations of the octahedra. The second mode, associated with rotations of octahedra, is lost upon transition to the tetragonal phase. Such a sequence of phase transitions is characteristic of most APs with $m = 2$, studied to date, such as $Bi_2SrTa_2O_9$, $Bi_{2.1}Sr_{0.85}Ta_2O_9$, $Bi_2PbTa_2O_9$ and $Bi_2PbNb_2O_9$ [151, 254, 255, 509, 510]. Table 1.7 presents the temperatures of phase transitions for all these compounds.

However, there is a direct transition from the $A2_1am$ orthorhombic phase to the $I4/mmm$ tetragonal paraelectric phase without any signs of the existence of an intermediate $Amam$ orthorhombic phase in the $SrBi_2Nb_2O_9$ and $Bi_3Ti_{1.5}W_{0.5}O_9$ compounds [164, 371], as shown by diffraction studies. Their phase transition temperatures are 735°C and 440°C, respectively. Group theory calculations have shown that the direct transition from $A2_1am$ to $I4/mmm$ must be a first-order transition. For both compounds, there is a violation of the smooth course of the dependence of the cell parameters on temperature near the phase transition point. Based on this, it was concluded that the phase transitions can be attributed to first-order transitions.

At present, the question of what factors contribute to the stabilization of the intermediate structural phase remains open. The presence of an intermediate orthorhombic phase separating the ferroelectric phase $B2cb$ and the high-temperature phase $I4/mmm$ was also revealed in the AP $Bi_4Ti_3O_{12}$ with $m = 3$ [497]. Based on the consideration of the phase sequence from the viewpoint of group theory, it was concluded that the most probable sequence of

Table 1.7. Temperatures of phase transitions of APs with $m = 2$.

Compound	$A2_1am$, $a > b$	$A2_1am$, $a = b$	$Amam$, $a > b$	$Amam$, $a = b$	$I4/mmm$
$Bi_3Ti_{1.5}W_{0.5}O_9$	700	700	—	—	735
$SrBi_2Nb_2O_9$	400	400	—	—	440
$Bi_2SrTa_2O_9$	300	300	—	—	550
$Bi_{2.1}Sr_{0.85}Ta_2O_9$	375	—	375	—	550
$Bi_2PbNb_2O_9$	575	—	575	625	650
$Bi_2PbTa_2O_9$	200	200	—	400	500

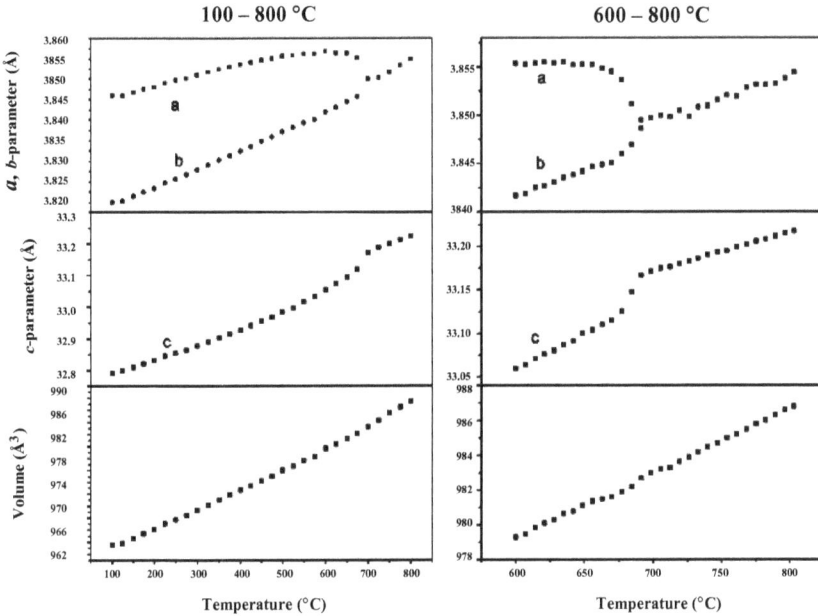

Fig. 1.20. Temperature dependence of the $Bi_4Ti_3O_{12}$ cell parameters.

phase transitions in this case is: $B2cb$ at 670°C → $Cmca$ at 695°C → $I4/mmm$. The temperature dependence of the cell parameters for this compound is shown in Figure 1.20. Weak signs of a slight orthorhombic distortion in the temperature range above the Curie point T_C (from 550°C to 650°C) were also found in $SrBi_4Ti_4O_{15}$ ($m = 4$) [153], while direct experimental evidence of the presence of intermediate phase in $BaBi_4Ti_4O_{15}$ could not be obtained [192].

1.10. Solid Solutions and Doping Effects

Doped materials and solid solutions, based on APs, are of considerable interest to scientists. This is explained by the fact that these materials often have improved ferroelectric characteristics compared to the original compounds. By varying the composition, it is possible to obtain desired dielectric characteristics for specific applications.

The simplest representative of compounds from the series of APs is the well-known ferroelectric Bi_2WO_6 with a Curie temperature T_C

of about 950°C. For this compound, $m = 1$ and the perovskite layers consist of an infinite two-dimensional sequence of vertex-connected WO_6 octahedra. It is a ferroelectric of the displacement type and has a rather complex structure at room temperature, characterized by rotations of the octahedrons, as well as atomic displacements. Modifications of this compound were also obtained and studied by replacing W with Nb and Ta [167]. Doping leads to the formation of oxygen vacancies in WO_6 octahedrons and thus the formula of the obtained compounds becomes $Bi_2W_{1-x}M_xO_{6-x/2}$, where M is the metal, $0 < x < 0.15$. Based on data on neutron diffraction from powder samples, the cell parameters of $Bi_2W_{0.9}Ta_{0.1}O_{6-x/2}$ were determined, and the parameters of Bi_2WO_6 were refined from $a = 5.4385(1)$ Å, $b = 16.4552(3)$ Å, $c = 5.4614(1)$ Å and $a = 5.4365(1)$ Å, $b = 16.4333(3)$ Å, $c = 5.4577(2)$ Å, respectively. A complete refinement of the structure of these compounds in the space group $Pca2_1$ was also carried out. Doped compounds $Bi_2W_{1-x}M_xO_{6-x/2}$ are of interest because they have ionic conductivity, so the greatest attention is paid to the mobility of anions. In particular, by means of neutron diffraction, a study was performed on the distribution of oxygen vacancies, which showed that the anions were uniformly distributed over equatorial and apial positions in oxygen octahedrons. Using numerical simulations, it was found that the lowest energy barrier to migration was localized along the edges of the octahedrons connecting the equatorial and apial oxygen positions.

The greatest interest was attracted to doped compounds, based on $SrBi_2Nb_2O_9$. In particular, materials with a substitution of Sr^{2+} by Ca^{2+} and Ba^{2+} (up to 30 at. %) [471] and La, Ca (up to 10 at. %) [106] in the *A*-positions, as well as a substitution of Nb^{5+} by V^{5+} (up to 30 at. %), Cr and Mo (up to 20 at. %) [279] in the *B*-positions. The substitution of Sr by Ca^{2+} and La^{2+} leads to an increase in the Curie temperature T_C, but the addition of Ba, on the contrary, leads to its decrease, with a linear dependence of the Curie temperature T_C on the impurity concentration. All three impurities in the Nb position lead to an increase in the Curie temperature T_C.

Doping has a significant effect on the dielectric characteristics of the material. Small additions of Ca and Ba (within 2.5 and 5 at. %, respectively) lead to a rather noticeable increase in the maximum permittivity, but a further increase in the impurity concentration leads to a sharp drop in the value of this maximum. The study of the permittivity of $SrBi_2Nb_2O_9$, doped with Mo and Cr, and its frequency dependence showed that, in the case of both types of doping, it increases with increasing concentration of impurity atoms, with the sharpest increase observed in the region of low concentrations.

Doping of $SrBi_2Nb_2O_9$ with Ca^{2+}, Ba^{2+} and V atoms leads to a decrease in the dielectric loss tangent and conductivity, while the La-doped material has a higher dielectric loss tangent than that of the initial compound.

Recently, the effect of doping on the properties of compounds has also been studied for compositions based, on Bi_3NbTiO_9, with a donor $Bi_3Nb_{1.2}Ti_{0.8}O_9$ doping and an acceptor $Bi_3Nb_{0.8}Ti_{1.2}O_9$ doping [475]. All these materials are ferroelectrics with a Curie temperature $T_C \approx 900°C$. $Bi_3Nb_{1.2}Ti_{0.8}O_9$ demonstrated an increase in dielectric constants and a decrease in the coercive field. At the same time, in contrast, in $Bi_3Nb_{0.8}Ti_{1.2}O_9$, a decrease in dielectric constants occurred with an increase in the coercive field, which is due to the influence of oxygen vacancies that prevent the movement of domain walls. The study of thermal depolarization showed that in the initial material, as well as in the material with the acceptor-type doping, a rapid decrease in piezoelectric parameter d_{33} is observed near the Curie temperature T_C, while in the case of donor-type doping, a significant decrease in d_{33} is observed in the temperature range well below the Curie point T_C. In another doped compound, $Sr_{1+x}Bi_{2-(2/3)x}(V_xTa_{1-x})_2O_9$ (where $x = 0.0, 0.1, 0.2$) [387], an increase in the Curie temperature T_C and a sharp increase in the dielectric constant are also observed in comparison with the initial material. Doping also leads to a decrease in the dielectric loss tangent. In doped samples, a frequency dispersion of the dielectric constant was also found, namely, the temperature, at which the dielectric constant passes through a maximum, shifts toward lower temperatures with increasing frequency.

Finally, the solid solutions of $Bi_{4-x}Ba_xTi_{3-x}Nb_xO_{12}$ (where $0 \leq x \leq 1.4$) were studied in Ref. [31]. Dielectric measurements have shown that when x increases, the properties of the material gradually acquire relaxor properties.

1.11. Conclusions

The attention of researchers has been drawn for a long time to perovskite-like compounds, which can be used to solve many practical and applied problems due to their numerous physical properties. Of particular interest are bismuth-containing layered ferroelectrics, first described by Aurivillius. Aurivillius phases are characterized by low permittivity, high Curie temperatures, and low aging rates; therefore, they have long been the main candidates for materials, used to creating nonvolatile memory microcircuits. The given properties of the Aurivillius phases are significantly dependent on the composition and structure of the compounds, therefore, the identification of the "composition–structure–properties" dependences makes it possible to control the properties of the material by varying its composition. Revealing the patterns of structural changes near the polymorphic transition from ferroelectric to paraelectric modification at the Curie temperature is necessary to establish its mechanism.

The preparation of previously unknown representatives of the Aurivillius phases, as well as the modification of known compounds, are necessary to search for new promising materials that are superior in properties to analogs. Therefore, establishing the relationships between the composition, structure and properties of the material is an urgent task.

Chapter 2

Analytical Approaches to Modeling Switching Kinetics of Ferroelectrics

2.1. Introduction

Ferroelectrics are a class of dielectrics with spontaneous polarization, the dipole moment of which can be reoriented by inducing an external electric field. The process of switching polarization in such materials is the result of the formation of self-similar structures, which results in a self-similar structure of domains and fractality of electrical responses.

It is known that Riesz kernels generalize the kernels of classical theory, and convolution with them implements negative fractional powers of the Laplace operator. Along with hypersingular integrals, operators of the Riesz potential type arise in new areas of analysis and its applications for solutions of various problems of mathematical physics have been explored. A fractals-based mathematical model of the switching process of ferroelectric polarization under the action of an electric field has been proposed in Ref. [270] and then developed in Ref. [281], which reported the results of its computational realization into a framework of a hybrid fractal-stochastic approach.

By mathematical modeling the kinetics of switching ferroelectrics in the injection mode, the corresponding mathematical models assume the solution of the initial-value problem for the equation of dynamics of the domain boundary of a ferroelectric with a fractional-order derivative. A Hardy–Littlewood-type theorem on the limitation of a potential with a power-logarithmic kernel and a Lebesgue-summable density with a specific power-law weight was proved. It was shown that for large orders of potential, the image of a function from this class belongs to a weighted generalized Hölder space with a power-logarithmic characteristic [87, 88].

On the whole, recent achievements of fractional calculations are found as numerous examples of applications in such areas as genetic algorithms, heat transfer, percolation, chemistry, irreversibility, condensed matter physics, mechanics of deformable solids, control systems and so on [3, 77, 253, 293, 404]. For example, in Ref. [293], an electromechanical system (whose operating principle is the inverse of the energy harvesting system) is considered with a fractional-order capacitor by using a fractional-order Duffing-quintic equation. The system dynamics and synchronization were studied because the authors concluded that the fractional-order component can strongly influence the performance of the system, especially in the definition of the route to chaos and the onset of synchronization. An energy harvesting system with fractional-order viscoelastic material was investigated in Ref. [52]. It was shown that the fractional-order property of the material increases high-energetic chaotic motion as well as inter-well periodic oscillation. In Ref. [214], it was supposed that the electric circuit had elements whose characteristics had fractional-order derivatives. At the same time, the elastic properties of the mechanical component were modeled by using fractional-order power law similarly to those in Refs. [72, 73]. The performance improvement or worsening of the energy harvesting mechanism was analyzed in relation to the effective contribution to this property of the order of fractional derivative and deflection. For bistable configuration, the bifurcation diagram of the fractional-order deflection (from intra-well to inter-well dynamics) was discussed in the cases of a pure harmonic excitation and broadband excitation. It was stated that the fractional-order derivative effectively improves the efficiency of the harvester only for a big amplitude of perturbation by using the hardening behavior of the system. Conversely, varying the fractional power deflection and, consequently, lowering the depth of the potential energy of the mechanical component promotes the greater oscillation amplitude, corresponding to the optimal power output [214].

The aim of this chapter is further development of the methods of fractional calculus (fractional differentiation and integration) with creation of promising analytical approaches to modeling the ferroelectrics switching kinetics.

2.2. Mathematical Model of Ferroelectric Domain Boundary

The proposed mathematical model includes an initial problem for the dynamic equation of the ferroelectric domain boundary, which is of the following general form:

$$\frac{d^\alpha u}{dt^\alpha} = F(t, u(t)), \quad u(t_0) = u_0, \tag{2.1}$$

where $u(t)$ is the unknown distance function, and $F(\cdot)$ symbolizes the known right-hand side of the equation, $t \in [t_0, T]$, $t_0 \geq 0$. The order of the differential operator is assumed to be fractional, that is $0 < \alpha < 1$, which reflects the fractal nature of the displacement of the domain wall. Depending on the type of initial condition, two definitions of the fractional derivative are used, namely, Riemann–Liouville and Caputo, coupled with certain expressions.

The purpose of the present study is to generalize the formulation of the aforementioned problem by considering variable-order operators of fractional integro-differentiation under condition (2.2):

$$0 < \alpha(t) < 1, \quad t \in [t_0, T], \tag{2.2}$$

which assumes the temporal localization of the dynamic features of the domain boundaries in a ferroelectric. Moreover, a mathematical formalization of dynamic properties is proposed, reflecting several important characteristics of the stated task:

(i) from the viewpoint of the physics of the polarization switching process (the local characteristics of the increment of the velocity of the domain boundaries);
(ii) from the viewpoint of the analytical properties of the function $u(t)$ (the description of its local smoothness properties);
(iii) from the viewpoint of the study of the integro-differential equation (the characteristics of its solvability and stability).

These features are expressed by a Hölder-type condition with a local modulus of continuity, as follows:

$$\sup_{|\tau|<\delta} |f(t+\tau) - f(t)| \le c\omega(x,\delta), \quad \delta \in [0, T-t], \quad t \in [t_0, T],$$

(2.3)

where t is considered the temporal variable, and the analytical properties of function $\omega(\cdot)$ define the assumptions of the considered model.

Thus, owing to the regularization of Equation (2.1) with the derivative of variable-order $\alpha(\cdot)$, it is possible to reduce it to the Fredholm integral equation of the second kind:

$$D_{t_0+}^{\alpha(\cdot)} I_{t_0+}^{\alpha(\cdot)} \varphi(t) = \varphi(t) + \int_{t_0}^{t} \kappa(t, t-\tau)\varphi(\tau)d\tau, \quad u = I_{t_0+}^{\alpha(\cdot)}\varphi, \quad (2.4)$$

where the fractional differentiation operator $D_{t_0+}^{\alpha(\cdot)}$ and fractional integration operator $I_{t_0+}^{\alpha(\cdot)}$ are defined further, and the kernel $\kappa(\cdot)$ has an explicit expression. Equation (2.4) can be solved with numerical methods, while condition (2.2) guarantees a qualified smoothness of the solution.

2.3. Mathematical Techniques of Fractional Dynamics

As it is stated in Ref. [412], the analysis of mappings defined on fractal sets, being a generalization of the analytical theories of smooth manifolds, proposes a variety of open issues, which are researched actively. Usually, the starting point of fractal analysis is the definition of the Laplacian on a fractal, which is no longer a differential operator in the common sense but is characterized by a number of desirable properties. The same is true for the operation of integration on a fractal that is no longer in the framework of the traditional definition. Nevertheless, it is very important to overcome the difficulties associated with mathematical formalism, since mathematical analysis on fractals and integration of fractal regions can be used to describe dynamic characteristics of objects modeled within a new scientific paradigm.

In this chapter, we define fractional integro-differentiation operators that are directly involved in the formulation of the problem of the dynamics of ferroelectric domain boundaries and consider the general role of fractional analysis operators in the description of complex environments and interactions. We shall consider some of the theoretical achievements in the field of such operators, focusing on the Hölder formalization of their smoothness. We intend to highlight issues not only from the "one-dimensional theory", but also to form an impression of the situation in the case of spaces of arbitrary dimension, as well as those relationships between these two areas that may be useful. Our description of the "multidimensional theory" will be based on the Riesz model of fractional integro-differentiation, the instruments of which are potentials and hypersingular integrals. Both these classes consist of integral operators, but for the former are of mild singularity, the order of which is smaller than the dimension of a space, on sets of which integration is defined. However, in a hypersingular integral, the order of singularity dominates the dimension of the integration space. Generalizing the properties of objects under consideration, we shall finally formulate the results on the Hölder smoothness of potentials and hypersingular integrals on the sets of an arbitrary metric space. As fractional integrals are related to the Riesz potentials in a rather simple manner, many of the relevant theoretical results can be derived from these basic analytical statements.

The general motivation of this chapter is to provide analytical approaches and solid theoretical results, which could remain unnoticed by specialists in fractal models of mathematical physics but can be useful for developing this new field of study in the way of qualitative characterization of the research objects. An encyclopedic presentation of mathematical models in the fields of fractional dynamics and corresponding issues is provided by Ref. [412]; Refs. [49, 50] are also of interest in this connection.

2.3.1. *Operators of fractional calculus*

Let us consider the variable-order left-sided fractional integral:

$$I_{t_0+}^{\alpha(\cdot)} f(t) := \frac{1}{\Gamma[\alpha(t)]} \int_{t_0}^{t} \frac{f(\tau)}{(t-\tau)^{1-\alpha(t)}} d\tau, \quad t > t_0, \qquad (2.5)$$

along with the left-sided operator of fractional differentiation:

$$D_{t_0+}^{\alpha(\cdot)} f(t) := \frac{f(t)}{\Gamma[1-\alpha(t)]} (t-t_0)^{-\alpha(t)}$$

$$+ \frac{\alpha(t)}{\Gamma[1-\alpha(t)]} \int_{t_0}^{t} \frac{f(t)-f(\tau)}{(t-\tau)^{1+\alpha(t)}} d\tau, \qquad (2.6)$$

both defined under condition (2.2). Hereinafter, $\Gamma(\cdot)$ denotes the gamma-function.

Let it be noted that the fractional derivative $D_{t_0+}^{\alpha(\cdot)}$ demands a function f to have some "nice" analytical properties: for instance, to be from the variable exponent Hölder space $H^{\lambda(t)}([t_0,T])$ with $\lambda(t) > \alpha(t)$ for all $t \in [t_0,T]$. Another important point is that the term "left-sided" implies the existence of a way to determine the "right-sided" analogs of these operators. To maintain the clarity of the presentation, we assume they are defined through the so-called "reflection operator" Q:

$$Q\varphi(t) := \varphi(t_0+T-t), \quad t \in [t_0,T], \qquad (2.7)$$

which implements the following mapping of integrals [346, p. 34]:

$$I_{T-}^{\alpha(\cdot)} Q = Q I_{t_0+}^{\alpha(\cdot)}, \quad Q I_{T-}^{\alpha(\cdot)} = I_{t_0+}^{\alpha(\cdot)} Q, \qquad (2.8)$$

that can be considered as the conceptual definition of the right-sided fractional integral. For more details, we refer to Ref. [349].

The history of such operators begins with the works of Abel, Liouville, Riemann, Holmgren and other famous mathematicians of the 19th century, who tried to explain the meaning of fractional calculus operations through illustrative mathematical problems and analytical facts. Thus, in the case of constant order α, $0 < \alpha < 1$, the prototype of operator (2.5) is known as the left-sided fractional Riemann−Liouville integral, defined for every function integrable on (t_0,T) in the Lebesgue sense. It arises naturally from considering Abel equation for a tautochrone problem.

At the same time, (2.6) is known, in the case of a constant order, as an analog of the Marchaud derivative on the segment of the real axis, while it is originally considered as a more convenient form of Liouville fractional derivative on the real axis $\mathbb{R}^1$ [349, pp. 109, 225].

The Riemann–Liouville fractional derivative, which is defined as the following form:

$$\mathcal{D}_{t_0+}^{\alpha}f(t) := \frac{1}{\Gamma(1-\alpha)}\left[\frac{f(t_0)}{(t-t_0)^{\alpha}} + \int_{t_0}^{t}\frac{f'(\tau)}{(t-\tau)^{\alpha}}d\tau\right] \tag{2.9}$$

relates to the Marchaud fractional derivatives owing to rather simple transforms but differs from it by the domain of definition: Liouville's definition demands a function to be differentiable, while the definition by Marchaud does not. It can be stated also that Marchaud fractional derivatives allow more freedom at infinity [349, p. 111].

It is important for applications that fractional integration and differentiation defined above are reciprocal operations for a suitable set of functions. Hereinafter, we shall use the following symbols:

- $C = C(\Omega)$ — for the space of continuous functions, and
- $C^1 = C^1(\Omega)$ — for the differentiable ones, defined on a set Ω.

Thus, $C^1([t_0, T])$ is the domain of definition for the Riemann–Liouville fractional derivative (2.9).

Let us consider the space $L^p = L^p(t_0, T)$ of p-integrable functions:

$$L^p([t_0, T]) := \left\{f : \int_{t_0}^{T}|f(\tau)|^p d\tau\right\}, \quad 1 \le p < \infty. \tag{2.10}$$

We shall call functions from $L^1([t_0, T])$ just "integrable" and specify the Lebesgue exponent if needed. Let $I_{t_0+}^{\alpha}(L^p)$, $\alpha > 0$, define the space of functions $f(t)$ that are represented by the left-sided fractional integral of order α with a summable function:

$$f = I_{t_0+}^{\alpha}\varphi, \quad \varphi \in L^p(t_0, T), \quad 1 \le p < \infty.$$

The following theorem is known from Ref. [349, p. 44]:

Theorem 2.1. *Let $\alpha > 0$. Then the equality*

$$\mathcal{D}_{t_0+}^{\alpha}I_{t_0+}^{\alpha}\varphi = \varphi(t) \tag{2.11}$$

is valid for any integrable function $\varphi(x)$, while

$$I_{t_0+}^{\alpha}\mathcal{D}_{t_0+}^{\alpha}f = f(t) \tag{2.12}$$

is satisfied for $f \in I_{t_0+}^{\alpha}(L^1)$.

The same is true for the constant order Marchaud fractional derivative, but in the case of variable-order fractional calculus operators (2.5) and (2.6), the result of Theorem 2.1 does not hold. Instead, the following statement on the regularization was proved in Ref. [346] for the real-axis operators

$$
I_+^{\alpha(\cdot)} f(t) := \frac{1}{\Gamma(t)} \int_{-\infty}^t \frac{f(\tau)}{(t-\tau)^{1-\alpha(t)}} d\tau, \quad \alpha(t) > 0,
$$

$$
D_+^{\alpha(\cdot)} f(t) := \frac{\alpha(t)}{\Gamma[1-\alpha(t)]} \int_0^\infty \frac{f(t) - f(t-\tau)}{\tau^{1+\alpha(t)}} d\tau, \quad 0 < \alpha(t) < 1.
$$

$$(2.13)$$

Theorem 2.2. *Let*

$$
0 < \alpha(t) < 1, \quad t \in (-\infty, T), \quad T \le \infty, \tag{2.14}
$$

satisfy the following conditions:

$$
\alpha \in C^1(-\infty, T), \quad \sup_{t \in (-\infty, T)} |\alpha'(t)| < \infty. \tag{2.15}
$$

Then for sufficiently good functions, the following representation holds:

$$
D_+^{\alpha(\cdot)} I_+^{\alpha(\cdot)} f(t) = (I + K)f := f(t) + \int_{-\infty}^t \kappa(t, t-\tau) f(\tau) d\tau,
$$

$$(2.16)$$

where the operator K is defined by the following kernel:

$$
\kappa(t, \xi) = \frac{\alpha(t) \sin[\pi \alpha(t)]}{\pi} \int_0^\xi \frac{(\xi-\tau)^{\alpha(t)-1} - \theta(\xi, t, \tau)}{\tau^{1+\alpha(t)}} d\tau,
$$

$$
\theta(\xi, t, \tau) := \frac{(\xi-\tau)^{\alpha(t-\tau)-1} \Gamma[\alpha(t)]}{\Gamma[\alpha(t-\tau)]}.
$$

$$(2.17)$$

Only later, a result, based on Theorem 2.2, was obtained for operators (2.5) and (2.6) in the generalized variable Hölder spaces [432]. This result is discussed below and proposed for use in this chapter to generalize the problem that is considered in the next section.

2.3.2. *Fractional derivatives applications*

It has been outlined, therefore, that there are multiple ways to define a fractional derivative. These definitions appear to be equivalent for some sets of "nice" functions or are related to each other owing to specific transformations. Another important definition is the Caputo fractional derivative that is defined as

$$^{C}D_{t_0+}^{\alpha} f(t) := \frac{1}{\Gamma(1 - \{\alpha\})} \int_{t_0}^{t} \frac{d^{[\alpha]+1} f(\tau)}{d\tau^{[\alpha]+1}} \frac{d\tau}{(t - \tau)^{\{\alpha\}}}, \quad \alpha > 0,$$

(2.18)

which demands the function f to be differentiable in the sense of absolutely continuous functions [349, pp. 2–3]. The brackets in (2.18) symbolize the integer part $[\alpha]$ and the fractional part $\{\alpha\}$ of the real α.

In general, choosing the right form of a fractional derivative depends on the correctness of the problem statement: initial, boundary conditions, or both, if intended. Thus, the Caputo derivative is used to consider the problem (2.1) in the case of an inhomogeneous initial condition, while the Riemann–Liouville fractional derivative is applied if $u_0 \equiv 0$. Nevertheless, they are related to each other owing to the following identity, which is true in the considered case of $0 < \alpha < 1$ [353]:

$$^{C}D_{t_0+}^{\alpha} u(t) = \mathcal{D}_{t_0+}^{\alpha} u(t) - r_0^{\alpha}(t) u_0, \quad r_0^{\alpha}(t) := \frac{t^{-\alpha}}{\Gamma(1 - \alpha)}.$$

(2.19)

Another issue is with regards to the calculation process and the preferences in the calculating algorithm [201].

As it is noticed in Ref. [54], with references to Refs. [5, 83, 96], the Marchaud definition of fractional derivative provides a clear mechanical interpretation of the fractional operators in the models of non-local continuums, for instance, solids that are characterized by non-local interactions. The "non-local" characteristic means that the result of some physical interaction in a region of a body, which can be modeled by a fractal subset, is effected by the response over the whole structure. It is remarkable that in modern physics both local and non-local models are increasingly carried out using non-standard fractional derivatives.

2.3.3. *Fractional integral on fractal sets*

In contrast to fractional derivatives, the definition of a fractional integral is more specific. In this section, the role of this operator is outlined, according to the materials of Ref. [412].

Let Ω symbolize a non-empty subset of the n-dimensional Euclidian space R^n, and let $\{E_i\}$ be a countable family of disjoint subsets of diameter at most ε, covering Ω, as follows:

$$\forall i \quad \Omega \subset \bigcup_{i=1}^{\infty} E_i, \quad \operatorname{diam}(E_i) \leq \varepsilon, \tag{2.20}$$

where

$$\operatorname{diam}(E_i) = \sup_{x,y \in E_i} r(x,y), \tag{2.21}$$

$r(\cdot)$ is the distance. The characteristic function of a subset E_i is defined as follows:

$$\chi_{E_i}(x) = \begin{cases} 1, & x \in E_i; \\ 0, & x \notin E_i. \end{cases} \tag{2.22}$$

A function $f(x)$, defined on Ω, is called a simple one, if

$$f(x) = \begin{cases} f_k, & x \in E_k; \\ 0, & x \notin \bigcup_{k=1}^{\infty} E_k, \end{cases} \quad f(x) = \sum_{i=1}^{\infty} f_i \chi_{E_i}(x), \quad f_i \in R. \tag{2.23}$$

Let the simple function $f(x)$ be integrable on Ω, that is

$$\forall i : f_i \neq 0 \quad \mu_H(E_i) < \infty, \tag{2.24}$$

where the measure is defined in terms of a fractional dimension α, as follows:

$$\mu_H(\Omega, \alpha) = \lim_{\varepsilon \to 0} \mu_H^{(\varepsilon)}(\Omega, \alpha),$$

$$\mu_H^{(\varepsilon)}(\Omega, \alpha) = \inf \left\{ \sum_{i=1}^{\infty} \omega(\alpha)[\operatorname{diam}(E_i)]^{\alpha} : \Omega \subset \bigcup_{i=1}^{\infty} E_i, \operatorname{diam}(E_i) \leq \varepsilon \right\},$$

$$\omega(\alpha) = \frac{\pi^{\alpha/2}}{2^n \Gamma(\alpha/2 + 1)}. \tag{2.25}$$

Then the Lebesgue integral can be defined as

$$\int_\Omega f(x)d\mu_H(x) = \sum_{i=1}^\infty f_i\mu_H(E_i,\alpha)$$

$$= \omega(\alpha) \lim_{\text{diam}(E_i)\to 0} \sum_{E_i} f(x_i)[\text{diam}(E_i)]^\alpha. \quad (2.26)$$

Since it is always possible to divide the subset $\Omega \subset \mathbb{R}^n$ into the collection of ε-cubes, defined as

$$E_{i_1\cdots i_n} = \{(x_1,\ldots,x_n) \in \Omega : \quad i_k\varepsilon \le x_k \le (i_k+1)\varepsilon\}, \quad (2.27)$$

where i_k are naturals, then

$$d\mu_B(x) = \lim_{\text{diam}(E_{i_1\cdots i_n})\to 0} \omega(\alpha)[\text{diam}(E_i)]^\alpha, \quad (2.28)$$

where

$$\mu_B(\Omega,\alpha) = \omega(\alpha)\zeta^\alpha(\Omega), \quad \zeta^\alpha(\Omega) = \lim_{\varepsilon\to 0} \zeta_\varepsilon^\alpha(\Omega), \quad (2.29)$$

$$\zeta_\varepsilon^\alpha(\Omega) = \inf \left\{ \sum_{i=1}^\infty [\text{diam}(E_{i_1\cdots i_n})]^\alpha : \Omega \subset \bigcup_{i_1\cdots i_n}^\infty E_{i_1\cdots i_n}, \right.$$

$$\left. \text{diam}(E_{i_1\cdots i_n}) \le \varepsilon\sqrt{n} \right\}.$$

As Ω is a subset of the Euclidean space, it can be parameterized using polar coordinates: $\rho = r(x,0) = |x|$ and the angle φ. Within this parametrization, a spherically symmetric covering by the $E_{\rho,\Omega}$ sets around the center at the origin can be considered, thus

$$d\mu_B(\rho,\varphi) = \lim_{\text{diam}(E_{r,\varphi})\to 0} \omega(\alpha)[\text{diam}(E_{\rho,\varphi})]^\alpha = d\varphi^{\alpha-1}\rho^{\alpha-1}dr.$$

$$(2.30)$$

If $f(x)$ is assumed symmetric with respect to some center point $x_0 \in \Omega$, that is $f(x) = \text{const}$ for $\forall x \in \Omega : |x - x_0| = \rho$ and arbitrary ρ, then the center of symmetry can be shifted with the following transformation:

$$\Omega \to \Omega' : x \to x' = x - x_0, \tag{2.31}$$

and the integral over Ω can be represented as

$$\int_\Omega f d\mu_B = \frac{2\pi^{\alpha/2}}{\Gamma(\alpha/2)} \int_0^\infty f(\rho)\rho^{\alpha-1} dr = \frac{2\pi^{\alpha/2}\Gamma(\alpha)}{\Gamma(\alpha/2)}(I_-^\alpha f)(0), \tag{2.32}$$

where

$$(I_-^\alpha f)(x) := \frac{1}{\Gamma(D)} \int_x^\infty \frac{f(t)}{(t-x)^{1-\alpha}} dt, \quad -\infty < x < \infty, \tag{2.33}$$

is the right-sided Riemann–Liouville integral, defined on the real half-axis. Along with I_-^α, the left-sided fractional integral is considered as [349, p. 33]

$$(I_+^\alpha f)(x) := \frac{1}{\Gamma(\alpha)} \int_{-\infty}^x \frac{f(t)}{(x-t)^{1-\alpha}} dt, \quad -\infty < x < \infty. \tag{2.34}$$

Fractional integration operators with constant limits occur in many fields of mathematics, and owing to analogies in mathematical physics, such operators are called potential-type operators. Thus, the Riesz potential over the real axis is considered as the following operator [349, p. 214]:

$$(I^\alpha f)(x) = \frac{1}{2\Gamma(\alpha)\cos(\alpha\pi/2)} \int_{-\infty}^\infty \frac{f(t)}{|t-x|^{1-\alpha}} dt,$$
$$\mathrm{Re}\,\alpha > 0, \quad \alpha \neq 1, 3, 5, \ldots, \tag{2.35}$$

and its useful modification is

$$(H^\alpha f)(x) = \frac{1}{2\Gamma(\alpha)\sin(\alpha\pi/2)} \int_{-\infty}^\infty \frac{\mathrm{sgn}(x-t)}{|x-t|^{1-\alpha}} f(t) dt, \quad \mathrm{Re}\,\alpha > 0, \tag{2.36}$$

where $\alpha \neq 2, 4, 6, \ldots$, with I^α and H^α defined on functions $f \in L_p(\mathbb{R})$, $1 \leq p < 1/\mathrm{Re}\,\alpha$, while $0 < \mathrm{Re}\,\alpha < 1$. Obviously, there is a connection between these potentials and the fractional integrals [349, p. 214]:

$$I^\alpha = [2\cos(\alpha\pi/2)]^{-1}(I_+^\alpha + I_-^\alpha),$$
$$H^\alpha = [2\sin(\alpha\pi/2)]^{-1}(I_+^\alpha - I_-^\alpha) \tag{2.37}$$

defined on the functions from $L^p(\mathbf{R}^1)$ and bounded, in the case of $1 < p < 1/\operatorname{Re}\alpha$, from $L^p(\mathbf{R}^1)$ into $L^q(\mathbf{R}^1)$ with $q = p(1 - p\operatorname{Re}\alpha)^{-1}$.

2.4. Fractional-dynamic Problem Statement

The conceptual formulation of the problem is based on Refs. [270, 281], in which the process of switching the ferroelectric polarization was described from the viewpoint of mathematical modeling of fractal systems. The starting observation of the validity of this approach was the fact that the polarization switching process is the result of the formation of self-similar structures, so that the domain configurations of many ferroelectrics are characterized by a self-similar structure, and electrical responses are fractal patterns.

After some transformations, the authors [281] formulate the following problem for the fractional differential equation of the dynamics of the domain boundary in dimensionless coordinates:

$$\frac{d^\alpha s}{dw^\alpha} = \exp\left(-\frac{\tau_2}{\tau_1 w + \frac{\tau_3}{2l}(1 + \cos^2\varphi)\frac{Ls}{2}}\right), \quad s(w_0) = 0, \quad (2.38)$$

where $s = x/L$ is the dimensionless distance, satisfying the following inequality:

$$0 \le s(w) \le 2(1 - l/L), \quad (2.39)$$

a dimensionless time is found as

$$w = t/\tau_1, \quad w \in [t_0/\tau_1, T/\tau_2], \quad (2.40)$$

where $\tau_1 = L/v_\infty$ is the characteristic run-time of the crystal thickness by the domain wall at a speed of v_∞; $\tau_2 = \varepsilon\varepsilon_0 E_2 L/(jl)$ is the time of charge accumulation that creates a field E_2 at current density j, ε is the dielectric permittivity of the sample, ε_0 is a dielectric constant; $\tau_3 = 2P_S/j$ is the time during which the current j initiates the appearance of charges with a surface density $2P_S$; $\cos^2\varphi = y^2/(x^2 + y^2)$.

The purpose of this study is to obtain analytical results relevant to the study of the problem (2.18) in the case of a common statement as

$$D_{t_0+}^{\alpha(\cdot)} u(t) = R(u, t), \quad u(t_0) = u_0, \quad (2.41)$$

when the operators (2.6), defined accordingly, are considered as a fractional derivative.

2.4.1. *Hölder continuity*

It is known that the classical Hölder condition arises as a natural generalization of the concept of continuity, defined in the context of equations of mathematical physics. In their classical work [123], David Gilbarg and Neil Trudinger, by a systematic presentation of the general theory of elliptic equations of the second order, selected the Laplace equation and its inhomogeneous form of the Poisson equation, as the main model. They are present in the following form, respectively:

$$\Delta u = 0, \quad \Delta u = f, \tag{2.42}$$

where Δ is the Laplace operator.

The initial point of the study is the Gauss–Ostrogradsky formula, the consequence of which is Green's identity connecting integration over a region Ω of Euclidean space R^n with integration over its boundary $\partial\Omega$

Let us fix a certain point $y \in \Omega$, then the normalized fundamental solution $G(\cdot)$ of the Laplace equation has the form [123, p. 18]

$$G(x-y) = G(|x-y|) = \begin{cases} \left[\dfrac{2(2-n)\pi^{n/2}}{\Gamma(n/2)}\right]^{-1} |x-y|^{2-n}, & n > 2, \\ \dfrac{1}{2\pi}\ln|x-y|, & n = 2. \end{cases} \tag{2.43}$$

By using Green's representation formula, we have the following equation:

$$u(y) = \int_{\partial\Omega} \left(u\frac{\partial G}{\partial v}(x-y) - G(x-y)\frac{\partial u}{\partial v}\right) ds$$

$$+ \int_{\Omega} G(x-y)\Delta u\, dx, \quad y \in \Omega, \tag{2.44}$$

For every integrable function f, the integral

$$w(x) = \int_\Omega G(x - y)f(y)dy \tag{2.45}$$

is the Newtonian potential with density f. An important feature is the fact that only the continuity of the function does not guarantee the twice continuous differentiability of the Newtonian potential. To correct this circumstance, let us require continuity in terms of the Hölder condition.

Let us select a point $x_0 \in \mathrm{R}^n$ and a bounded set $D(x_0)$, containing x_0. If the function f is defined on $D(x_0)$, then it satisfies the Hölder condition with an exponent λ at the point x_0 if the following inequality is carried out:

$$[f]_{\lambda;x_0} := \sup_{x \in D(x_0)} \frac{|f(x) - f(x_0)|}{|x - x_0|^\lambda} < \infty, \quad 0 < \lambda < 1, \tag{2.46}$$

which entails the continuity of the function f at x_0 also in terms of the class of continuous functions C. In the case of $\lambda = 1$, the function f is called Lipschitz continuous. If we assume that D is an arbitrary, not necessarily bounded, set, then f is called uniformly Hölder continuous with exponent λ in D if the following condition is satisfied:

$$[f]_{\lambda;D} := \sup_{\substack{x,y \in D \\ x \neq y}} \frac{|f(x) - f(y)|}{|x - y|^\lambda} < \infty, \quad 0 < \lambda \leq 1, \tag{2.47}$$

and it is locally Hölder continuous with exponent λ in D if f is uniformly Hölder continuous with exponent λ on compact subsets of D [123, p. 52].

Finally, we have the following result [123, p. 55]:

Theorem 2.3. *Let f be bounded and locally Hölder continuous with exponent $\lambda \leq 1$ in the bounded domain $\Omega \subset \mathrm{R}^n$, and let w be the Newtonian potential with density f. Then $w \in C^2(\Omega)$ and $\Delta w = f$.*

Thus, the introduction of the concept of Hölder continuity is motivated by the questions of the smoothness of the Newtonian potential and, as a consequence, the question of the unambiguous solvability of the classical Dirichlet problem: $\Delta u = f$ in Ω, $u = \varphi$ on $\partial\Omega$, discussed in detail in Ref. [123, p. 56].

The class of functions continuous by Hölder within the framework of the condition (2.26) will be denoted by the symbol $H^\lambda = H^\lambda(D)$. The value on the left-hand side of the inequalities (2.25) and (2.26) is called the λ-Hölder factor of f at x_0 or in D, respectively. It is easy to show that the elements of H^λ form a Banach space, the norm in which can be defined in terms of this coefficient. Let us now consider how classes H^λ are generalized to reflect the continuity property in a more flexible way.

2.4.2. *Generalized Hölder spaces*

Obviously, the condition (2.26) for λ-factor is equivalent to the following requirement:

$$|f(x) - f(y)| \leq c|x - y|^\lambda, \quad 0 < c < \infty, \tag{2.48}$$

for any values of $x, y \in D$, $x \neq y$. As a result of this circumstance, it is possible to restrict the oscillation of the function f in D by any arbitrary function $\omega(\cdot)$ of the distance between the points x and y.

Returning from the abstract formulation of the question to the case of one-dimensional segments relevant to the physical problem, we agree that the space H^λ is defined by the norm

$$\|f\|_{H^\lambda[t_0,T]} = \|f\|_{C[t_0,T]} + \sup_{t_0 \leq t_1, t_2 \leq T} \frac{|f(t_1) - f(t_2)|}{|t_1 - t_2|^\lambda}. \tag{2.49}$$

We shall define a similar space H^ω by using the concept of a continuity modulus. As a result, a new concept of Hölder continuity will be proposed, which is of interest in the framework of the stated problem.

Let us give a function $f(t)$ that is continuous on the segment $[t_0, T]$ as an element of the class C. The continuity modulus of function $f(t)$ means the following function of a real variable δ:

$$\omega_f(\delta) := \sup_{\substack{|t_1-t_2|\leq\delta \\ t_0 \leq t_1, t_2 \leq T}} |f(t_1) - f(t_2)|, \quad 0 < \delta \leq T - t_0. \tag{2.50}$$

The continuity modulus of a continuous function has a set of properties that are of a defining nature. Namely, we will call an arbitrary function $w(\delta)$ a function of the modulus of continuity type if it has the following set of properties:

(i) $w(0) = 0$;
(ii) $w(\delta)$ is a non-decreasing function of δ;
(iii) $w(\delta)$ is semi-additive, that is: $w(\delta_1 + \delta_2) \le w(\delta_1) + w(\delta_2)$;
(iv) $w(\delta)$ is continuous.

Thus, we intend to majorize the continuity module (2.46) by a function with similar properties. Considering the specifics of Hölder spaces, two more requirements are added to the above conditions, forming a functional class [134, p. 57].

We assume that $w(\delta) \in \Phi$, $\delta \in (0, T - t_0]$, if $w(\delta)$ is the function of the continuity module type, and

$$\delta \int_0^{T-t_0} \frac{w(\tau)}{\tau(\delta + \tau)} d\tau \le c w(\delta), \quad 0 < c < \infty. \qquad (2.51)$$

Moreover, the equivalence condition ($\sim$) should be fulfilled:

$$w'(\delta) \sim \frac{w(\delta)}{\delta}. \qquad (2.52)$$

Finally, let $w \in \Phi$. We will say that a function $f(t)$ belongs to a class $H^w = H^w[t_0, T]$ if it is defined on $[t_0, T]$ and satisfies the condition

$$[f]_{w;[t_0,T]} := \sup_{0 < \delta \le T - t_0} \frac{w_f(\delta)}{w(\delta)} < \infty. \qquad (2.53)$$

The function $w(\delta)$, $\delta \in (0, T - t_0]$, is called the characteristic function of space H^w or, briefly, its characteristic. By introducing in H^w the norm

$$\|f\|_{H^w} = \|f\|_{C[0, T - t_0]} + [f]_{w;[t_0,T]}, \qquad (2.54)$$

it transforms in the Banach space of functions, which we will call the generalized Hölder space. Obviously, this space extends the notion of uniform continuity by Hölder, while the space H^λ is just a particular case of this new definition in the case of a power modulus of continuity:

$$w(\delta) = \delta^\lambda, \quad 0 < \lambda \le 1, \quad \delta \in (0, T - t_0]. \qquad (2.55)$$

Finally, together with the spaces of generalized Hölder type, we consider the spaces $H^{\omega(\cdot)} = H^{\omega(\cdot)}[t_0, T]$, which can be found within the generalization of the condition (2.49).

Let the function $f(t)$ be defined and bounded on the segment $[t_0, T]$. By analogy with ω_f, we will call the following function a local module of the submetry:

$$M_f(\delta, t) := \sup_{|h| < \delta} |f(t + h) - f(t)|, \quad \delta \in [0, T - t_0], \quad t \in [t_0, T].$$

$$(2.56)$$

The submetry of metric spaces is a mapping in which the image of every ball is a ball having the same radius. This concept is most fully revealed when considering arbitrary metric spaces. However, in our particular case, it is important to note that the local module of the submetry is not a function of the type of the module of continuity, since its semi-additivity, generally speaking, cannot be guaranteed.

In order to preserve required generality of the presentation, the following approach can considered. For a given function $f(t)$, we will call a function $M_f(\delta, t)$ by a local continuity module if it is a function of the type of continuity module, or alternatively its minimal majorant from this class. Moreover, it can be shown that $M_f(\delta, t)$ is equivalent to the local modulus of continuity as the supremum of the oscillations of a function, defined on a system of nested segments containing a given point t.

Anyway, we are interested in a condition of the Hölder type, formulated in terms of $M_f(\delta, t)$. Namely, we denote the space of functions $f \in C[t_0, T]$ through $H^{\omega(\cdot)} = H^{\omega(\cdot)}[t_0, T]$ such that:

$$M_f(\delta, t) \le c\omega(\delta, t), \quad \delta \in (0, T - t_0], \quad t \in [t_0, T], \quad c > 0,$$

$$(2.57)$$

where the function $\omega(\cdot)$ is now a function of two real variables capable of reflecting correctly the local smoothness characteristics of the given function $f(t)$. At the same time, the norm in the space $H^{\omega(\cdot)}$ can be introduced similarly to those used earlier:

$$\|f\|_{H^{\omega(\cdot)}} = \|f\|_{C[t_0,T]} + \sup_{\substack{t\in[t_0,T]\\ \delta>0}} \frac{M_f(\delta,t)}{\omega(\delta,t)}. \tag{2.58}$$

By assuming $\omega(\delta,t) = \delta^{\lambda(t)}$, the variable exponent Hölder space $H^{\lambda(\cdot)}$ can be introduced.

By $H_0^{\omega(\cdot)} = H_0^{\omega(\cdot)}[t_0,T]$, we shall denote the subspace of $H^{\omega(\cdot)}$ whose elements are evaluated to zero at the point t_0.

Finally, we will need an analog of the class Φ, adapted to the specification of the characteristics of the new type. For these purposes, in the theory of $H^{\omega(\cdot)}$-spaces it is suggested to use the generalized Zygmund–Bary–Stechkin classes:

$$\Phi_{\beta(\cdot)}^{\delta(\cdot)}, \quad 0 \leq \delta(t) < \beta(t), \quad t \in [t_0,T], \tag{2.59}$$

assuming that, as a function of δ, $\omega(\delta,t)$ is continuous and almost increasing on segment $[0, T - t_0]$ uniformly for $t \in [t_0,T]$. Moreover,

$$\forall t \in [t_0,T] \quad \lim_{\delta\to+0} \omega(\delta,t) = 0 \tag{2.60}$$

and there are integral estimations, as follows:

$$\int_0^h \left(\frac{h}{\rho}\right)^{\delta(t)} \frac{\omega(\rho,t)}{\rho} d\rho \leq c\omega(h,t), \quad \int_h^{T-t_0} \left(\frac{h}{\rho}\right)^{\beta(t)} \frac{\omega(\rho,t)}{\rho} d\rho \leq c\omega(h,t), \tag{2.61}$$

where $0 < h < T - t_0$, and the constant $c > 0$ does not depend on h and t.

2.5. Review of Some Results

The first result on fractional calculus in Hölder spaces was carried out by Weyl [459], who showed that periodic functions, which satisfy the Hölder condition with the exponent λ, have continuous fractional derivatives of order $\alpha < \lambda$. For non-periodic functions, a similar result was proved by Montel [278], while Hardy and Littlewood managed to describe the action of fractional integrals and derivatives on the "classical" Hölder spaces H^λ in a more specific way [147].

In particular, they proved a theorem on the acting of the constant-order fractional integral (2.5) from the Hölder space of functions, which are zero at the point t_0, to the space of the same kind with the exponents being changed from λ to $\lambda + \alpha < 1$, while the constant-order fractional derivative lowers the order of Hölder continuity, which is expressed by the transform of λ into $\lambda - \alpha$. In Ref. [339], similar results on the constant-order fractional integrals were carried out in the case of the constant exponent Hölder spaces with weights of the exponential type.

The fractional calculus operators in function spaces with variable characteristics have been of some interest over the recent years. Thus, for the one-dimensional case, corresponding theoretical problems were firstly considered in Refs. [124, 190] for the variable exponent Hölder spaces, and, in Ref. [347], for the spaces $L^{p(\cdot)}$. Noticeable results here include the theorems on isomorphisms of the variable exponent Hölder spaces by the fractional integral operator.

The variable-order fractional calculus operators were proposed in Ref. [350] and the first results on their Hölder continuity were obtained in Ref. [338]. In Ref. [430], they were considered in the generalized variable Hölder spaces for the first time, while the main results on the topic, on which the present chapter is based, were presented in Refs. [431, 432]. Reference [298], which can be recommended as one of the basics in this field of study, is another important paper on the fractional calculus of variable order.

The topic of fractional calculus on sets of multidimensional spaces presents a subsequent layer of the whole theory. It needs to be discussed in the framework of this review as it outlines specific features of the generalized Hölder continuity that are out of scope in the one-dimensional case.

To be specific, let Euclidean space R^n of vectors with real coordinates be considered, $\mathrm{R} = \mathrm{R}^1$, in which

$$x \cdot y = \sum_{i=1}^{n} x_i\, y_i, \quad |x - y| = \sqrt{\sum_{i=1}^{n} (x_i - y_i)^2} \qquad (2.62)$$

define, respectively, the inner product and the distance between two points $x = (x_1, \ldots, x_n)$, $y = (y_1, \ldots, y_n)$, $x, y \in \mathrm{R}^n$.

One of the main models of fractional calculus on functions of many variables is the Riesz fractional integro-differentiation, which is carried out in terms potential-type operators and hypersingular integrals defined on R^n [349, p. 483]. These two kinds of operators are used to implement fractional powers of the Laplace operator, as well as to represent differential operators in partial derivatives [349, p. 527]. The Riesz potential-type operators can be represented as

$$I_\Omega^{\alpha,\nu} f(x) := c_\alpha \int_\Omega \frac{f(y)}{|x-y|^{\mu(\alpha,n)}} \ln^\nu \frac{\rho(x,y)}{|x-y|} dy, \quad x \in \Omega, \quad \operatorname{Re}\alpha > 0,$$

(2.63)

where $\nu, c_\alpha \in R$, $\rho(\cdot)$ is a given function, which we shall assume to be identically equal to the unit by default. The classical Riesz potentials are derived from (2.63) within the following parameters:

$$\Omega = R^{n-1}, \quad \mu(\alpha,n) = n-1-\alpha, \quad \nu \in \{0,1\}, \quad c_\alpha = \frac{1}{\gamma_n(\alpha)},$$

(2.64)

where the normalizing constant is given as

$$\gamma_n(\alpha) = \begin{cases} 2^\alpha \pi^{n/2} \Gamma\left(\frac{\alpha}{2}\right)/\Gamma\left(\frac{n-\alpha}{2}\right), & \begin{aligned} &\alpha \neq n+2k, \\ &\alpha \neq -2k; \end{aligned} \\[2ex] 1, & \alpha = -2k; \\[2ex] (-1)^{(n-\alpha)/2} \pi^{n/2} 2^{\alpha-1}\left(\frac{\alpha-n}{2}\right)!\Gamma\left(\frac{\alpha}{2}\right), & \alpha = n+2k, \end{cases}$$

(2.65)

with integer k. It is worth mentioning that these integral operators also attract attention because their kernels include the ones of the classical theory (Newtonian and logarithmic kernels) and Green's kernels associated with a region as special and limiting cases [224, p. 43]. However, while the mapping properties of these potentials on $\Omega = R^n$ for the functions from L^p are described by the Sobolev theorem, their action is not easy to consider on elements of the Hölder spaces due to the unboundedness of the integration set. Indeed, as the Hölder-type conditions require the absolute value of the difference between two neighbor function values be considered, it is important to take infinity into account. The same holds also for the problem on an arbitrary bounded set, where the effect of a boundary must be considered. At the same time, a convenient approach for the Hölder continuous functions of many variables can be derived from the spherical analog of Riesz fractional calculus.

A unit sphere in R^n is the following set of points equidistant from the origin:

$$S^{n-1} := \{x \in R^n : |x| = 1\}, \quad |x| = \sqrt{\sum_{i=1}^{n} x_i^2}. \quad (2.66)$$

The spherical Riesz potential was firstly introduced in Ref. [345] as

$$K^\alpha f(x') := \left(I^\alpha \frac{f(t')}{|t|^\alpha} \right)(x), \quad 0 < \alpha < n,$$

$$x' := \frac{x}{|x|}, \quad x \in R^{n-1}, \quad x' \in S^{n-1}. \quad (2.67)$$

The divergent integral here is realized in the sense of the following generalized functions:

$$\langle K^\alpha f, \varphi \rangle = \langle |x|^{-\alpha} f(x'), I^\alpha \varphi \rangle, \quad \varphi \in \Phi(R^{n-1}), \quad (2.68)$$

where Φ is the space of those and only those Schwartzian functions that are orthogonal to polynomials [349, p. 487]. The operation of (2.67) is understood as spherical Riesz integration.

It is clear that S^{n-1} can be mapped onto R^{n-1}. For $n = 2$, this correspondence can be established, for instance, by the mapping: $t \to (\cos t, \sin t)$, $t \in R$, while for multidimensional spaces there is a mapping by the stereographic projection that matches each point $\xi \in S^{n-1}$ with $x \in \dot{R}^n := R^n \cup \{\infty\}$:

$$\xi_k = \frac{2x_k}{|x|^2 + 1}, \quad k = \overline{1, n}; \quad \xi_{n+1} = \frac{|x|^2 - 1}{|x|^2 + 1}. \quad (2.69)$$

In this connection, the distances and measure elements are transformed in the following way [276, p. 207]:

$$|\xi - \sigma| = \frac{2|x - y|}{\sqrt{|x|^2 + 1}\sqrt{|y|^2 + 1}}, \quad d\sigma = 2^{n-1}(|y|^2 + 1)^{1-n}dy. \quad (2.70)$$

Owing to this fact, the results for the Riesz potential as a fractional calculus operator in the analysis of functions of many real variables can be derived from the results on spherical potentials, which are, at the same time, of much interest.

Many popular models of mathematical physics are carried out on spheres, or hyperspheres in the case of high-dimensional vector spaces. This contributed to the motivation for creating the fractional calculus on a sphere, analogous to the Riesz fractional integro-differentiation. The first definitions and basic facts from this theory appeared in Ref. [345]. It turned out that a meaningful definition of the Riesz potential as the integral operator on a unit (hyper)sphere preserves the structure, given by (2.63), but at the same time, owing to the relation

$$|\xi - \sigma| = \sqrt{2}\sqrt{1 - \xi \cdot \sigma}, \quad \xi, \sigma \in S^{n-1} \tag{2.71}$$

translates the issue into the plane of a new theoretical approach, based on the concept of spherical convolution operators. These operators, generally defined as

$$Kf(\xi) = \int_{S^n} k(\xi \cdot \sigma) f(\sigma) d\sigma, \quad \xi \in S^n \tag{2.72}$$

are suitable for studying from the viewpoint of spectral theories. Indeed, the representation of any operator in the form of (2.72) as a series of spherical harmonics, being unique, can characterize this operator by the coefficients of such a representation [348, p. 10]. The sequence of these coefficients is called the Fourier–Laplace multiplier of (2.72).

A fundamentally relevant question is about the conditions for the continuity of the spherical convolution operator and, moreover, its smoothness: namely, in what terms can the concept be defined and by what methods can we investigate the existence of this operator and in what way can these terms be related to spectral properties. By now, a rich theory based on applying the concept of the Hölder spaces and Zygmund-type estimates has been developed to answer the first two basic analytical questions, while for the third one there is an approach based on the asymptotical properties of the Fourier–Laplace multiplier.

Specific Hölder spaces on a sphere were introduced in Ref. [458] as special cases of more generalized functional spaces, proposed also in Ref. [294]. Equivalent normalizations in the Hölder spaces on a sphere were studied in Refs. [351, 434]. The generalized Hölder spaces, defined on a sphere in terms of an arbitrary continuity modulus, were considered in Refs. [91, 92]. The spherical potential-type operator of a constant real order in the generalized Hölder spaces was considered in Ref. [422].

Potential-type operators defined in these two metric subspaces of R^n are elements of the wider class whose members are integral operators with mild singularities in their kernels. An overview of the basic results on the properties of operators (2.63) is provided by Ref. [433], considering not only the case of a specific characteristic of the potential-type operator, but its general property to have discontinuity of a uniform function. The topic of two polar spherical Riesz potential-type operators of the kind

$$K_{\pm}^{\alpha} f(x) = \int_{S^{n-1}} \frac{f(\sigma)}{|x - \sigma|^{n-1-\alpha}|x + \sigma|^{n-1-\alpha}} d\sigma, \quad x \in S^{n-1}, \quad \operatorname{Re}\alpha > 0$$

(2.73)

was also covered by this paper. However, providing many of the results became possible by introducing the class of Fourier–Laplace multipliers with a given asymptotic in Ref. [422]. Based on the properties of this class, the following embeddings were proved for a real α:

$$I_{S^{n-1}}^{\alpha,0}(H^{\omega}(S^{n-1})) \subseteq H^{\omega_{\alpha}}(S^{n-1}), \quad \omega_{\alpha}(t) := t^{\alpha}\omega(t),$$

$$I_{\dot{R}^{n-1}}^{\alpha,0}\left(H^{\omega(\cdot)}\left(\dot{R}^{n-1}, (|x|^2 + 1)^{\frac{n+\alpha-1}{2}}\right)\right)$$

$$\subseteq H^{\omega_{\alpha}}\left(\dot{R}^{n-1}, (|x|^2 + 1)^{\frac{n-\alpha-1}{2}}\right).$$

(2.74)

Moreover, the conditions of the isomorphism

$$I_{S^{n-1}}^{\alpha,0}(H^{\omega}(S^{n-1})) = H^{\omega_{\alpha}}(S^{n-1})$$

(2.75)

were proved in Ref. [422] for $\alpha \in R$ and, in Ref. [428], for a complex α. They are a particular case of a theorem proved in Ref. [459] for the spherical convolutions (2.1) and used here for investigating the solutions of (2.2) in the spaces H^ω.

The boundedness theorem for the potential-type operator with a power-logarithmic kernel, which stated the following embedding:

$$I_{S^{n-1}}^{\alpha,\nu}(H^\omega(S^{n-1})) \subseteq H^{\omega_{\alpha,\nu}}(S^{n-1}), \quad 0 < \mathrm{Re}\,\alpha < 1,$$

$$\omega_{\alpha,\nu}(t) := t^{\mathrm{Re}\,\alpha} \ln^\nu \tfrac{r}{t}\omega(t),$$

(2.76)

with arbitrary real value was carried out in Ref. [339] based on Zygmund-type estimates without analysis of the operator's spectral properties. This gap has been partially filled by Ref. [124], where the Fourier–Laplace multiplier is considered, and the result on the inverse operator is provided in the form of the following theorem:

Theorem 2.4. *The inverse operator for $I_{S^{n-1}}^{\alpha,1}$ is expected to be a composition of the spherical hypersingular integral*

$$D^\alpha f(x) = \frac{1}{\gamma(n,-\alpha)} \lim_{\varepsilon \to \infty} \int_{S_\varepsilon^{n-1}(x)} \frac{f(\sigma) - f(x)}{|x - \sigma|^{n-1+\alpha}} d\sigma, \quad x \in S^{n-1},$$

$$S_\varepsilon^{n-1}(x) := \{\sigma \in S^{n-1} : |\sigma - x| > \varepsilon\},$$

(2.77)

and a spherical convolution operator A with the specific multiplier, i.e.

$$(I_{S^{n-1}}^{\alpha,1})^{-1} = (c_\alpha E + D^\alpha)A, c_\alpha = \Gamma\left(\frac{n-1+\alpha}{2}\right)\Big/\Gamma\left(\frac{n-1-\alpha}{2}\right),$$

(2.78)

where E is the identity operator: $Ef = f$,

$$(Af)(x) = \int_{S^{n-1}} a(x \cdot \sigma)f(\sigma)d\sigma, \quad x \in S^{n-1},$$

$$a(\cdot) = \frac{1}{4}\pi^{-\frac{n}{2}}\Gamma\left(\frac{n-2}{2}\right)\sum_{m=0}^{\infty}(2m+n-2)l_m^{-1}C_m^{\frac{n-2}{2}}(\cdot),$$

(2.79)

and the kernel $a(\cdot)$ is generated by the Fourier–Laplace multiplier $\{l_m^{-1}\}_{m=0}^{\infty}$ with

$$l_m = -2^{\frac{n+\alpha-1}{2}} \pi^{\frac{n-1}{2}} \left(\frac{m}{m+n-3} \right)^{-1} \frac{(n-2)_m}{m\,((n-1)/2)_m}$$

$$\times \left(m + \frac{n-\alpha-1}{2} \right) \Gamma\left(\frac{\alpha}{2} \right)$$

$$\times \mathrm{B}^{-1}\left(\frac{n-1}{2}, m \right) \Gamma^{-1}\left(\frac{n-\alpha+1}{2} \right)$$

$$\times \left\{ 1 + \ln\frac{2}{r} \left[\psi\left(\frac{\alpha-n+3}{2} \right) + \psi\left(\frac{\alpha}{2} \right) \right.\right.$$

$$\left.\left. -\psi\left(\frac{\alpha-n+3}{2} - m \right) - \psi\left(\frac{\alpha+n-1}{2} + m \right) \right] \right\}. \qquad (2.80)$$

Here, $\Gamma(z)$, $\mathrm{Re}\,z > 0$, is the gamma-function, and the following special functions are used:

$$\mathrm{B}(z, w) = \frac{\Gamma(z)\Gamma(w)}{\Gamma(z+w)}, \quad \psi(z) = \frac{\Gamma'(z)}{\Gamma(z)} = \frac{d}{dz}\ln\Gamma(z),$$

$$(m)_k = m \cdot (m+1) \cdots\cdots (m+k-1), \quad k = 1, 2, 3, \ldots; \qquad (2.81)$$

$$\binom{m}{k} = \frac{(m)_k}{k!},$$

$C_m^{\lambda}(\cdot)$ denotes the Gegenbauer polynomials. Theorem 2.1 allows to formulate the following isomorphism, based on (2.5):

$$I_{S^n}^{\alpha,1}(H^{\omega}(S^n)) = H^{\omega_{\alpha},1}(S^n), \qquad (2.82)$$

where the characteristic $\omega_{\alpha,1}(\cdot)$ is derived from (2.6).

In Ref. [190], the acting of $I_{S^{n-1}}^{\alpha,\nu}$ from the space $L^p(S^{n-1})$ of p-integrable functions to the space $H^{\omega}(S^{n-1})$ with an arbitrary value of ν and $r > 2$ in (2.4) was studied, resulted as the embedding

$$I_{S^n}^{\alpha,\nu}(L^p(S^n)) \subseteq H^{\omega}(S^n), \quad \omega(t) = t^{\mathrm{Re}\,\alpha-\frac{n-1}{p}}\ln^{\nu}\frac{r}{t}. \qquad (2.83)$$

Finally, Ref. [347] provided the results, analogous to (2.6)–(2.8), in the case of weighted generalized Hölder spaces and the operator $I_{\dot{R}^{n-1}}^{\alpha,\nu}$

defined with the following parameters:

$$c_\alpha = 2^{\alpha-2\nu}\Gamma\left(\frac{n-1+\alpha}{2}\right)\Big/\Gamma\left(\frac{n-1-\alpha}{2}\right),$$

$$\rho(x,\sigma) = \frac{r}{2}w_\Pi(x)w_\Pi(\sigma), \quad r > 0,$$
(2.84)

where $\alpha \in R$ and

$$w_\Pi(x) := 1 + |x|^2, \quad x \in \dot{R}^{n-1},$$
(2.85)

is the weight function due to the stereographic projection. Specifically, the following embeddings were carried out:

$$I^{\alpha,\nu}_{\dot{R}^{n-1}}\left(H^{\omega_\alpha}\left(\dot{R}^{n-1}, w_\Pi^{-\frac{n+\alpha-1}{2}}\right)\right) \subseteq H^{\omega_{\alpha,\nu}}\left(\dot{R}^{n-1}, w_\Pi^{\frac{n-\alpha-1}{2}}\right),$$

$$I^{\alpha,\nu}_{\dot{R}^{n-1}}(L^p(\dot{R}^{n-1}, w_0)) \subseteq H^\omega\left(\dot{R}^{n-1}, w_\Pi^{\frac{n-\alpha-1}{2}}\right),$$
(2.86)

with the weight function

$$w_0(x) := 2^{p(n-2)}[w_\Pi(x)]^{\frac{(n-1)(2p-1)-\alpha}{2}-p},$$
(2.87)

where $\omega_{\alpha,\nu}$ and ω are given in (2.6) and (2.8). The first of the embeddings is an isomorphism for $\nu \in \{0, 1\}$.

The spherical Riesz potential of a constant complex order was studied in works of Shankishvili and the Zygmund-type estimates were carried out. In Ref. [345], the result was extended to the case of the spatial Riesz potential. In Ref. [423], the spherical convolution operator with specific asymptotics of its multiplicator was considered, and the conditions for it, bounded in the generalized Hölder spaces, were studied. The Zygmund-type estimates and smoothness properties of the spherical convolution operator of the potential type with an exponentially logarithmic kernel were studied in Ref. [429], and the theorems on boundedness were proven.

The spherical potential-type operator in the weighted generalized variable Hölder spaces was studied in Ref. [424], where the results on the isomorphism were obtained.

The spherical potential-type operator of a constant order with weak singularities on the poles was considered in Ref. [427], and the theorems on isomorphism were developed. In Ref, [344], the spherical variable-order Riesz potential was studied in the weighted and non-weighted generalized variable Hölder spaces, and the theorems on boundedness were proven.

2.6. Constant-Order Models

The first trivial results on the Equation (2.1) can be derived from Ref. [349, p. 51]. We shall consider it in the form

$$D_{t_0+}^{\alpha} f(t) = g(t), \quad t \in [t_0, T], \quad 0 < \alpha < 1, \tag{2.88}$$

and, by assuming $g \in L^1([t_0, T])$, express the solution as $f = I_{t_0+}^{\alpha} g$. We intend to provide a closed form of f.

Let it be recalled that a point t is said to be a Lebesgue point of a function $f \in L^1([t_0, T])$ if

$$\lim_{x \to 0} \frac{1}{x} \int_0^x [f(t - \tau) - f(t)] d\tau = 0. \tag{2.89}$$

Almost every point $t \in [t_0, T]$ is a Lebesgue point of $\in L^1([t_0, T])$.

Let $b(\tau)$ be a bounded function such that

$$|b(\tau)| < \varepsilon, \quad 0 < \tau < \delta(\varepsilon). \tag{2.90}$$

By considering an auxiliary function

$$H(\tau) := \tau[g(t) - b(\tau)], \tag{2.91}$$

the closed-form solution of (2.1) can be expressed as follows in terms of $b(\tau)$:

$$f(t) = g(t) + \frac{H(t - t_0)}{\Gamma(\alpha)(t - t_0)^{1-\alpha}} + g(t_0) \left[\frac{1-\alpha}{\alpha\Gamma(\alpha)} (t - t_0)^{\alpha} - 1 \right]$$

$$+ \frac{1-\alpha}{\Gamma(\alpha)} \int_0^{\delta} \tau^{\alpha-1} b(\tau) d\tau + \frac{1-\alpha}{\Gamma(\alpha)} \int_0^{t-t_0} \tau^{\alpha-1} b(\tau) d\tau. \tag{2.92}$$

It is interesting to note, that if $g \in H_0^\omega([t_0, T])$, then the local continuity modulus of g at the point t_0 can be used as a function b.

2.6.1. *The variable-order fractional calculus*

The aim of this paragraph is to propose valuable analytical approaches in the case of the variable fractional order. To be specific, we consider the general equation (2.1) in the form

$$D_{t_0+}^{\alpha(\cdot)} f = g, \tag{2.93}$$

by assuming that Equation (2.18) is a certain case that can be considered in the framework of previously stated connections between the fractional derivatives. We shall start from formulating a theorem, which is a direct analog of the theorem relevant to the generalized Hölder spaces. In conclusion, the spaces of variable-order integral operators will be defined.

We have the following theorem.

Theorem 2.5. *Let the following conditions hold:*

$$0 < \inf_{t \in [t_0, T]} \alpha(t) \leq \sup_{t \in [t_0, T]} \alpha(t) < 1, \tag{2.94}$$

$$\alpha \in C^1([t_0, T]), \quad \sup_{t \in [t_0, T]} |\alpha'(t)| < \infty, \tag{2.95}$$

$$w(\delta, t) \in M \cap \Phi_{1-\alpha(t)}^{\alpha(t)}, \quad \delta^{-\alpha(t)} w(\delta, t) \in M, \tag{2.96}$$

$$w(\delta, t + \delta) \leq c w(\delta, t), \quad \delta > 0, \quad c > 0, \tag{2.97}$$

where the constant c does not depend on t and δ. Then the following identity is true:

$$D_{t_0+}^{\alpha(\cdot)} I_{t_0+}^{\alpha(\cdot)} \varphi = (I + L)\varphi, \quad \varphi \in H_0^{\omega(\cdot)}([t_0, T]), \tag{2.98}$$

where L is a completely continuous operator on $H_0^{\omega(\cdot)}([t_0, T])$ that performs the following mapping:

$$L\varphi(t) = \int_{t_0}^t \kappa(t, t - \tau)\varphi(\tau)d\tau + \frac{\sin[\pi\alpha(t)]}{\pi(t - t_0)^{\alpha(t)}} \int_{t_0}^t \frac{\varphi(\tau)d\tau}{(t - \tau)^{1-\alpha(t)}}, \tag{2.99}$$

where $\kappa(t, \cdot)$ is the kernel from (2.99):

$$\kappa(t, t - \tau) = \frac{1}{\pi} \alpha(t) \sin[\pi \alpha(t)] \int_{t_0}^{t} (t - \tau - \xi)^{\alpha(t)-1}$$

$$- \frac{\Gamma[\alpha(t)]}{\Gamma[\alpha(t - \xi)]} (t - \tau - \xi)^{\alpha(t-\xi)-1} \frac{d\xi}{\xi^{1+\alpha(t)}}. \qquad (2.100)$$

The solution of (2.1) can be constructed as the variable-order fractional integral of a function φ, which is the solution of the Fredholm integral equation of the second kind:

$$\varphi(t) = g(t) - L\varphi(t), \quad t \in [t_0, T], \qquad (2.101)$$

for which numerical and semi-analytical methods can be applied [26].

2.6.2. *Spaces of bounded Riesz potential-type operators of variable order*

Let $\mathcal{V}$ and $\mathcal{W}$ be some vector spaces, then the set of linear operators of the kind $L : \mathcal{V} \to \mathcal{W}$ is a vector space, defined by the following relations:

$$(L_1 + L_2)f = L_1 f + L_2 f, \quad f \in \mathcal{V},$$
$$(\alpha L_1)f = \alpha L_1 f, \quad \alpha \in \mathbb{R}, \quad L_1, L_2 : \mathcal{V} \to \mathcal{W}, \qquad (2.102)$$

with the zero operator, usually denoted just as 0. This vector space of linear operators is denoted as $\mathcal{L}(\mathcal{V}, \mathcal{W})$. Assuming $\mathcal{V}$ and $\mathcal{W}$ are Banach spaces, and all the operators from $\mathcal{L}(\mathcal{V}, \mathcal{W})$ are bounded, $\mathcal{L}(\mathcal{V}, \mathcal{W})$ is to be considered as a normed vector space, equipped with the operator norm

$$\|L\| := \sup_{\substack{f \in D(L) \\ f \neq 0}} = \frac{\|Lf\|_{\mathcal{W}}}{\|f\|_{\mathcal{V}}} = \sup_{\substack{f \in D(L) \\ \|f\|_{\mathcal{V}}=1}} \|Lf\|_{\mathcal{W}}, \qquad (2.103)$$

and turns out to be a Banach space itself (see Theorem 3.5.5 in Ref. [162, p. 83]). By the symbol $D(L)$, the domain of L is denoted in the definition (2.1), and further we assume $D(L) = \mathcal{V}$. We shall also use this denotation for the domains of functions considered.

According to the results of Ref. [344], the Banach space of variable-order spherical Riesz potentials, $\varepsilon_\Omega[\alpha(x)] = n - 1 - \alpha(x)$, defined as

$$\mathcal{J}_{\mathbb{S}^{n-1}}^{\alpha(\cdot)}(\mathcal{H}^{\omega(\cdot)}(\mathbb{S}^{n-1}), \mathcal{H}^{\omega_\alpha(\cdot)}(\mathbb{S}^{n-1}, \alpha)), \quad \omega_\alpha(x,t) := t^{\operatorname{Re}\alpha(x)}\omega(x,t),$$
$$(2.104)$$

exists, if

$$\max_{x \in \mathbb{S}^{n-1}} \operatorname{Re}\alpha(x) < 1, \quad \omega \in \Phi_{1-\operatorname{Re}\alpha(x)}, \quad \text{and}$$

$$\operatorname{Re}\alpha(x) \geq 0, \quad \alpha \in \operatorname{Lip}(\mathbb{S}^{n-1}), \quad \max_{x \in \mathbb{S}^{n-1}} |\arg\alpha(x)| < \frac{\pi}{2} - \xi, \quad \xi > 0,$$
$$(2.105)$$

or

$$\min_{x \in \mathbb{S}^{n-1}} \operatorname{Re}\alpha(x) > 0, \quad M_d(\alpha, x, t) \leq ct^{\operatorname{Re}\alpha(x)}\omega(x,t), \quad c > 0.$$
$$(2.106)$$

As it is shown in detail in Ref. [426], the Banach space of variable-order Riesz potentials, bounded between generalized Hölder spaces with special weights, can be defined as the image of $\mathcal{J}_{\mathbb{S}^{n-1}}^{\alpha(\cdot)}(\mathcal{V}, \mathcal{W})$ by the stereographic projection operator Q. This operator is an isomorphism between the vector space $\{f \colon D(f) = \mathbb{S}^{n-1}\}$ and $\{\tilde{f} \colon D(\tilde{f}) = \dot{\mathbb{R}}^{n-1}\}$, where $\dot{\mathbb{R}}^{n-1}$ is the compactification of the hyperplane in $\mathbb{R}^n$ by the unique infinite point. Thus, the Banach space

$$\mathcal{J}_{\dot{\mathbb{R}}^{n-1}}^{\tilde{\alpha}(\cdot)}(\mathcal{H}^{\tilde{\omega}(\cdot)}(\dot{\mathbb{R}}^{n-1}), \mathcal{H}^{\tilde{\omega}_\alpha(\cdot)}(\dot{\mathbb{R}}^{n-1}, w_0\tilde{\alpha}))$$
$$= Q[\mathcal{J}_{\mathbb{S}^{n-1}}^{\alpha(\cdot)}(\mathcal{H}^{\omega(\cdot)}(\mathbb{S}^{n-1}), \mathcal{H}^{\omega(\cdot)}(\mathbb{S}^{n-1}, \alpha))] \qquad (2.107)$$

exists under analogous conditions if its elements are potential-type operators with a specific characteristic $c_0(\cdot)$ imposed, together with the weight function w_0, by the operator Q:

$$c_0(x, y) = [1 + d^2(y, 0)]^{-\frac{n+\alpha(x)}{2}},$$
$$w_0(x) = 2^{\alpha(x)}[1 + d^2(x, 0)]^{\frac{n-\alpha(x)}{2}}. \qquad (2.108)$$

Let $\mathcal{L}^p(\mathbb{S}^{n-1})$ denote the space of p-integrable functions. Based on the results of Ref. [459], it can be stated that the Banach space

$$\mathcal{J}_{\mathbb{S}^{n-1}}^{\alpha(\cdot)}(\mathcal{L}^p(\mathbb{S}^{n-1}), \mathcal{H}^{\lambda(\cdot)}(\mathbb{S}^{n-1})), \quad \lambda(x) = \alpha(x) - (n-1)/p$$

(2.109)

exists under the following conditions:

$$\alpha \in \text{Lip}(\mathbb{S}^{n-1}), \quad \min_{x \in \mathbb{S}^{n-1}} \text{Re}\,\alpha(x) > \frac{n-1}{p},$$

$$\max_{x \in \mathbb{S}^{n-1}} \text{Re}\,\alpha(x) < \frac{n-1}{p} + 1.$$

(2.110)

Similarly, reconsidered for the case of $\tilde{\alpha} = Q\alpha$, they provide the existence of the following Banach space:

$$\mathcal{J}_{\dot{\mathbb{R}}^{n-1}}^{\tilde{\alpha}}(\mathcal{L}^p(\dot{\mathbb{R}}^{n-1}, w), \mathcal{H}^{\tilde{\lambda}(\cdot)}(\dot{\mathbb{R}}^{n-1}, w_0))$$

$$= Q[\mathcal{J}_{\mathbb{S}^{n-1}}^{\alpha}(\mathcal{L}^p(\mathbb{S}^{n-1}), \mathcal{H}^{\lambda(\cdot)}(\mathbb{S}^{n-1}))],$$

$$w(x) = 2^{p(n-1)}[d^2(x, 0) + 1]^{p(1-n)}.$$

(2.111)

Finally, let Ω be an open bounded set in a metric space X, such that

$$\mu B(x, r) \leq K r^N, \quad r \to 0, \quad K > 0,$$

(2.112)

where $B(x, r) := \{y \in X : d(x, y) < r\}$, and $N > 0$ is not necessarily an integer. Further, Refs. [343, 425] prove the existence of the Banach space

$$\mathcal{J}_{\Omega}^{\alpha(\cdot)}(\mathcal{H}^{\omega(\cdot)}(\Omega), \mathcal{H}^{\omega_\alpha(\cdot)}(\Omega, \alpha)), \quad \varepsilon_\Omega[\alpha(x)] = N - \alpha(x).$$

(2.113)

2.7. Conclusions

This chapter studied fractional integro-differentiation operators that are directly involved in the formulation of the dynamical problem of displacement of the ferroelectric domain boundaries. The general role of fractional analysis operators was considered in the description of complex environments and interactions. Some theoretical achievements in the field of such operators were presented, focusing on the Hölder formalism of their smoothness. The issues not only regarding the "one-dimensional theory", but also the impressions of the situation in the case of spaces of arbitrary dimensions were high-lighted. Moreover, the useful relationships between these two areas were obtained, too. The description of the "multidimensional theory" was based on the Riesz model of fractional integro-differentiation, the instruments of which were potentials and hypersingular integrals. Both these classes consisted of integral operators, but for former, they were of mild singularity, the order of which was smaller than the dimension of a space on the sets of which integration was defined. However, in a hypersingular integral, the order of singularity dominated the dimension of the integration space. Finally, by generalizing the properties of objects under consideration, the results on the Hölder smoothness of potentials and hypersingular integrals on the sets of an arbitrary metric space were formulated. Since, the fractional integrals are related to the Riesz potentials in a rather simple manner, many of the relevant theoretical results can be derived from these basic analytical statements.

Chapter 3

Polarization Modeling and Application of Inhomogeneously Polarized Transducers in Acoustic Waveguides

3.1. Introduction

The vast majority of manufactured piezoceramic elements have simple shapes of, for example, rods, bars, round and rectangular plates, rings, and cylinders. Basically, they have uniform polarization, in the sense that the direction and magnitude of the remnant polarization vector are constant in the volume. However, converters with non-uniform polarization in the bulk are of interest for practical use. Transducers with this polarization may exhibit unusual properties and have original physical characteristics. In order to study piezoceramic elements with inhomogeneous polarization, it is necessary to involve models of polarization and deformation of such elements in strong electric and mechanical fields. According to the field of residual polarization and residual deformation, one can determine the physical modules of the material and, finally, involve analytical and numerical methods for solving problems, in which non-uniformly polarized piezoceramic elements excite or receive acoustic waves. This work is devoted to such problems. First, the main regularities of the physical characteristics of ceramics with inhomogeneous polarization are considered. Then the problem of longitudinal vibrations of a transversely polarized rod is considered. Next, we solve the problem for a mechanical system consisting of an acoustic path with a source and receiver of oscillations in the presence of impedance attenuation, in which piezoceramic elements have a uniform polarization. Finally, a problem is considered, similar to the previous one, but the piezoceramic transducers of the source and receiver of vibrations have a non-uniform polarization.

3.2. Constitutive Relations for Inhomogeneous Polarization

At the final stage of preparation of piezoceramic elements, the process of electric polarization is carried out: a large voltage is applied to the electrodes; the resulting electric field orients the spontaneous polarization vectors. After that, the ceramics acquire piezoelectric properties. The arrangement of electrodes on the surface of one or another transducer for carrying out such polarization can be arbitrary, and, therefore, non-uniform electric fields can arise inside the volume both in intensity and in direction. As a result, it is possible to obtain samples with inhomogeneous residual polarization, for which the elastic, piezoelectric and dielectric properties will depend both on the direction of the residual polarization vector and on the intensity of this polarization. However, along with the residual polarization in the samples, there is also a residual deformation, which also affects the indicated physical properties. The functional dependences of these properties on the residual parameters were studied in Ref. [364], where it was shown that the elastic and dielectric properties depend only on the residual deformation, while the piezoelectric properties depend only on the residual polarization. In the first case, the functional dependence is linear with respect to the residual strain tensor, and in the second case, it is linear with respect to the residual polarization vector. In the first case, it manifests itself weakly, and in the second case, in such a way that the absence of residual polarization leads to the absence of piezoelectric properties.

If we compare two piezoceramic samples with homogeneous and non-uniform polarization, we can find significant differences in their dynamic behavior, for example, in the difference of amplitude–frequency characteristics during harmonic oscillations. Therefore, the problem of studying the physical characteristics of a transducer with non-uniform polarization is relevant and of considerable interest, especially when such transducers are involved in a system with an acoustic path. This chapter is dedicated to this problem.

Let us present some results, concerning the physical properties of inhomogeneously polarized polycrystalline ferroelectric media. Under a representative volume, we mean a volume that is much smaller than the main volume but contains a huge number of spontaneous polarization vectors, and, as a result, spontaneous strain tensors. Let us define the residual polarization vector and the residual deformation tensor of this volume as the average characteristic of these elements:

$$\mathbf{P_0} = \frac{1}{N}\sum_{k=1}^{N}(\mathbf{p_s})_k \neq 0, \quad \varepsilon_0 = \frac{1}{N}\sum_{k=1}^{N}(\varepsilon_s)_k \neq 0. \tag{3.1}$$

It is known that both electric fields and mechanical stresses can rotate the residual polarization vectors and change the principal axes of the residual strain tensor when they reach threshold values. Both the electric field and mechanical stresses change both the remnant polarization and the remnant strain. However, it is impossible to cause residual polarization by mechanical stresses alone. The linear properties noted above relate to elastic compliances, piezoelectric moduli and dielectric permittivities, and can be written as

$$\mathbf{S}(\varepsilon_0) = \mathbf{S_0} + \mathbf{S_1} : \varepsilon_0,$$
$$\mathbf{d}(\mathbf{P_0}) = \mathbf{d_1} \cdot \mathbf{P_0}, \tag{3.2}$$
$$\mathbf{э}(\varepsilon_0) = \mathbf{э_0} + \mathbf{э_1} : \varepsilon_0.$$

where $\mathbf{S_0}$, $\mathbf{э_0}$ are the elastic compliance and permittivity tensors for the depolarized state, $\mathbf{S_1}$, $\mathbf{э_1}$, $\mathbf{d_1}$ are some additional tensors, the components of which are determined using auxiliary approaches. The appearance of the polarization vector leads to the appearance of piezoelectric properties in the ferroelectric material, while the residual deformation only slightly corrects the numerical values of the elastic compliances and permittivities. So in some works, for example, in Ref. [223], the effect of residual deformation is neglected. In Ref. [364], the constitutive relations for a representative volume in the coordinate axes, where z-axis coincides with the direction of the remnant polarization vector, are given in Voigt matrix representations. Physical characteristics are indicated by the same letters with "caps". For non-zero matrix components $\hat{S}_0$, $\hat{\jmath}_0$, we obtain the following expressions:

$$\hat{S}_{011} = \hat{S}_{022} = \hat{S}_{033} = \frac{1}{E}; \quad \hat{S}_{012} = \hat{S}_{013} = \hat{S}_{023} = -\frac{\nu}{E};$$

$$\hat{S}_{044} = \hat{S}_{055} = \hat{S}_{066} = \frac{2(1+\nu)}{E}; \quad \hat{\jmath}_{011} = \hat{\jmath}_{022} = \hat{\jmath}_{033} = \jmath_*. \tag{3.3}$$

To determine the matrices $\mathbf{S}_1$, $\jmath_1$, $\mathbf{d}_1$, the known values of physical moduli in the case of ceramics, polarized to saturation $\hat{\mathbf{S}}_{\text{sat}}$, $\hat{\mathbf{d}}_{\text{sat}}$, $\hat{\jmath}_{\text{sat}}$, are used, which have the forms

$$\hat{\mathbf{S}}_{\text{sat}} = \begin{pmatrix} \hat{S}_{11} & \hat{S}_{12} & \hat{S}_{13} & 0 & 0 & 0 \\ \hat{S}_{12} & \hat{S}_{11} & \hat{S}_{13} & 0 & 0 & 0 \\ \hat{S}_{13} & \hat{S}_{13} & \hat{S}_{33} & 0 & 0 & 0 \\ 0 & 0 & 0 & \hat{S}_{44} & 0 & 0 \\ 0 & 0 & 0 & 0 & \hat{S}_{44} & 0 \\ 0 & 0 & 0 & 0 & 0 & 2(\hat{S}_{11} - \hat{S}_{12}) \end{pmatrix},$$

$$\hat{\mathbf{d}}_{\text{sat}} = \begin{pmatrix} 0 & 0 & 0 & 0 & \hat{d}_{15} & 0 \\ 0 & 0 & 0 & \hat{d}_{15} & 0 & 0 \\ \hat{d}_{31} & \hat{d}_{31} & \hat{d}_{33} & 0 & 0 & 0 \end{pmatrix}, \tag{3.4}$$

$$\hat{\jmath}_{\text{sat}} = \begin{pmatrix} \hat{\jmath}_{11} & 0 & 0 \\ 0 & \hat{\jmath}_{11} & 0 \\ 0 & 0 & \hat{\jmath}_{33} \end{pmatrix}.$$

Then the final form of the constitutive relations according to Equations (3.2)–(3.4) is defined as follows:

$$\hat{S}_{\alpha\beta}(\varepsilon_0) = \hat{S}_{0\alpha\beta} + \frac{(\hat{S}_{\alpha\beta} - \hat{S}_{0\alpha\beta})}{\varepsilon_{\text{sat}}} \varepsilon_{0\,\text{III}}, \tag{3.5}$$

$$\hat{\jmath}_{mn}(\varepsilon_0) = \hat{\jmath}_{0\,mn} + \frac{(\hat{\jmath}_{mn} - \hat{\jmath}_{0\,mn})}{\varepsilon_{\text{sat}}} \varepsilon_{0\,\text{III}}, \tag{3.6}$$

$$\hat{d}_{m\alpha}(P_0) = \frac{|\mathbf{P}_0|}{p_{\text{sat}}} \hat{d}_{m\alpha}, \tag{3.7}$$

where $\varepsilon_{0\,\text{III}}$, $|\mathbf{P}_0|$ are the principal value of the tensile strain tensor and the modulus of the polarization vector, respectively. Relations

(3.5)–(3.7) completely determine the physical characteristics of the material in the irreversible process of deformation and polarization.

3.3. Longitudinal Oscillations of a Piezoceramic Rod with Inhomogeneous Transverse Polarization

3.3.1. *Classical problem statement*

Not many works have been devoted to the study of piezoceramic elements with inhomogeneous polarization, mainly in a one-dimensional formulation. As a rule, oscillations in the fundamental modes corresponding to the piezo constants e_{31} or e_{33} are considered [43, 44, 205, 363].

Let us first consider a rectangular rod with non-uniform polarization directed in the transverse direction, as shown in Figure 3.1.

We will assume that the left and right ends of the rod are subject to normal tension–compression loads, and the remaining faces are free from mechanical loads. The top and bottom electrodes are connected to a voltage generator, and the side end surfaces are non-electrodated.

$$\sigma_{12} = \sigma_{22} = \sigma_{23} = 0, \quad (y = \pm a/2),$$

$$\sigma_{13} = \sigma_{23} = \sigma_{33} = 0, \quad (z = \pm h/2),$$

$$\sigma_{11}(x, t) = \sigma e^{i\omega t}, \quad \sigma_{12} = \sigma_{13} = 0, \quad (x = \pm l), \qquad (3.8)$$

$$\varphi(x, t) = \pm \varphi_0 e^{i\omega t}, \quad \left(z = \pm \frac{h}{2}\right),$$

$$D_1 = 0, \quad (x = 0, \ x = l), \quad D_2 = 0, \quad (y = \pm a/2).$$

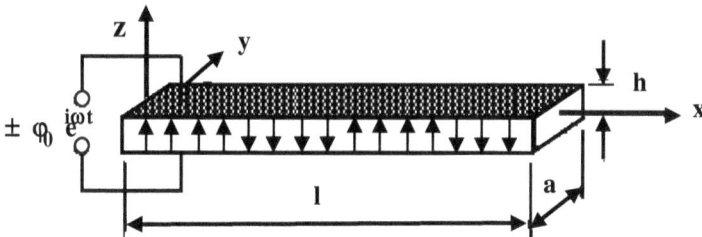

Fig. 3.1. Piezoceramic rod with transverse inhomogeneous polarization.

Based on the geometry of the rod and the type of boundary conditions, it is possible to apply the hypotheses of a plane stress state in the direction of the y-axis and z-axis. Then the problem is simplified, and we obtain the equations of the one-dimensional theory:

(i) the field relations:

$$\frac{\partial \sigma_{11}}{\partial x} = -\rho \omega^2 u, \quad \frac{\partial D_3}{\partial z} = 0, \tag{3.9}$$

(ii) the geometric relationships:

$$\varepsilon_{11} = \frac{\partial u}{\partial x}, \quad E_3 = \frac{2\varphi}{h}, \quad u = u(x), \quad \varphi = \varphi(z), \tag{3.10}$$

(iii) constitutive relations (3.5)–(3.7):

$$\varepsilon_{11} = S_{11}(\varepsilon_0)\sigma_{11} + d_{31}(p_0)E_3,$$
$$D_3 = d_{31}(p_0)\sigma_{11} + \vartheta_{33}(\varepsilon_0)E_3, \tag{3.11}$$

The transverse polarization of the rod is manifested by the fact that the physical moduli are the functions of residual deformation and residual polarization. Since these characteristics change inside the rod, the moduli are not constants but will be non-uniform functions of these quantities. It is easy to see that the residual polarization is directed along the z-axis. Therefore, the principal direction of the tensile strain tensor will also coincide with this axis. Further, this type of heterogeneity will be considered so the residual parameters do not change their direction along the z-axis, but change their value along the x-axis according to the law

$$p_0 = p_{\text{sat}} \sum_{m=1}^{M} \sin \frac{\pi m x}{l}, \quad \varepsilon_{0\,\text{III}} = \varepsilon_{\text{sat}} abs \left(\sum_{m=1}^{M} \sin \frac{\pi m x}{l} \right). \tag{3.12}$$

By regarding the deformation, it should be noted that for any direction of the polarization vector, it always remains a tensile deformation. Therefore, the sign of the absolute value is in Equation (3.12). According to Equations (3.5)–(3.7), physical moduli

η_{11}, d_{31}, $\mathfrak{z}_{33}$ take the following forms:

$$S_{11}(\varepsilon_0) = S_{011} + (S_{11} - S_{011})abs\left(\sum_{m=1}^{M} \sin\frac{\pi m x}{l}\right),$$

$$d_{31}(p_0) = d_{31} \sum_{m=1}^{M} \sin\frac{\pi m x}{l}, \qquad (3.13)$$

$$\mathfrak{z}_{33}(\varepsilon_0) = \mathfrak{z}_{033} + (\mathfrak{z}_{33} - \mathfrak{z}_{033})abs\left(\sum_{m=1}^{M} \sin\frac{\pi m x}{l}\right).$$

Relations (3.11) can be rewritten in another form:

$$\begin{aligned}
\sigma_{11} &= C(x)\varepsilon_{11} - e(x)E_3, \\
D_3 &= e(x)\varepsilon_{11} + \mathfrak{z}(x)E_3,
\end{aligned} \qquad (3.14)$$

where

$$C(x) = \frac{1}{S_{11}(\varepsilon_0)}, \quad e(x) = \frac{d_{31}(p_0)}{S_{11}(\varepsilon_0)}, \quad \mathfrak{z}(x) = \mathfrak{z}_{33}(\varepsilon_0) - \frac{d_{31}(p_0)}{S_{11}(\varepsilon_0)}. \qquad (3.15)$$

Relations (3.9), (3.10), (3.14) with boundary conditions (3.8) represent the classical formulation of the problem. Since the moduli of ceramics depend on the coordinates, we will build the solution of this problem by using the finite element (FE) method. To do this, we proceed to the generalized formulation of the problem by introducing a dimensionless coordinate $\xi = \frac{x}{l}$ and a dimensionless frequency $\Omega = \omega l\sqrt{\frac{\rho}{C_0}}$, $C_0 = \frac{1}{S_{011}}$.

3.3.2. *Weak problem statement*

Let us formulate a generalized statement of the problem. It is required to determine the function $u(\xi) \in W_2^1(0,1)$ satisfying the following integral equality:

$$\int_0^1 \frac{C(\xi)}{C_0}\frac{du}{d\xi}\frac{dv}{d\xi}d\xi - \Omega^2 \int_0^1 uv\,d\xi = \sigma(v(1) - v(0)) + \int_0^1 \frac{e(\xi)}{C_0}\frac{2\varphi_0}{h}\frac{dv}{d\xi}d\xi \qquad (3.16)$$

for any function $\forall v(\xi) \in W_2^1(0,1)$ satisfying the main boundary conditions (3.8).

We use one-dimensional three-node FE, for which we divide the segment $[0,1]$ into N identical elements, with length $\Delta = 1/N$. The extreme nodes of the k-th element have an uneven numbering, the internal nodes have an even numbering, and their coordinates are given by the following relations:

$$\xi_{2k-1} = \frac{k-1}{N}, \quad \xi_{2k} = \frac{2k-1}{2N}, \quad \xi_{2k+1} = \frac{k}{N}. \tag{3.17}$$

By using the replacement

$$\xi = \frac{\xi_{2k+1} - \xi_{2k-1}}{2}\zeta + \frac{\xi_{2k+1} + \xi_{2k-1}}{2} \tag{3.18}$$

let us rewrite the integral equality (3.16) in the form

$$\frac{1}{\Delta^2} \sum_{k=1}^{N} \int_{-1}^{1} \frac{C(\zeta)}{C_0} \frac{du}{d\zeta} \frac{dv}{d\zeta} d\zeta - \Omega^2 \sum_{k=1}^{N} \int_{-1}^{1} uv d\zeta$$

$$= \frac{\sigma}{\Delta}(v(1) - v(0)) + \frac{1}{\Delta} \sum_{k=1}^{N} \int_{-1}^{1} \frac{e(\zeta)}{C_0} \frac{2\varphi_0}{h} \frac{dv}{d\zeta} d\zeta. \tag{3.19}$$

By using a quadratic approximation:

$$u(\zeta) = [N_1(\zeta), N_2(\zeta), N_3(\zeta)] \begin{pmatrix} u_1 & u_2 & u_3 \end{pmatrix}^T \tag{3.20}$$

with local numbering of displacements in the nodes of the element and form functions:

$$N_1(\zeta) = \frac{\zeta^2 - \zeta}{2}, \quad N_2(\zeta) = \frac{-\zeta^2 + 1}{2}, \quad N_3(\zeta) = \frac{\zeta^2 + \zeta}{2}, \tag{3.21}$$

we obtain local matrices of stiffness, matrices of mass and a column of free terms, as follows:

$$C^{loc} = \frac{1}{\Delta^2} \int_{-1}^{1} \frac{C(\zeta)}{C_0} \begin{pmatrix} N_1' \\ N_2' \\ N_3' \end{pmatrix} (N_1' \ N_2' \ N_3') d\zeta,$$

$$M^{loc} = \int_{-1}^{1} \begin{pmatrix} N_1 \\ N_2 \\ N_3 \end{pmatrix} (N_1 \ N_2 \ N_3) d\zeta, \tag{3.22}$$

$$F^{loc} = \frac{2\varphi_0 e_0}{hC_0} \frac{1}{\Delta} \int_{-1}^{1} \frac{e(\zeta)}{e_0} \begin{pmatrix} N_1' \\ N_2' \\ N_3' \end{pmatrix} d\zeta.$$

To assemble general matrices, we notice that the corner elements of local matrices must be counted twice: first time from the previous finite element, and then from the next one. To take this circumstance into account, in Figure 3.2 the collected complete matrices were shown, in which the local matrices overlap each other precisely with the element that is counted twice. The complete system of linear algebraic equations has the form

$$(\mathbf{C} - \Omega^2 \mathbf{M}) \cdot \mathbf{U} = \mathbf{F} \tag{3.23}$$

Here, $\mathbf{U}$ is the vector of unknown nodal values, where C_0, e_0 are scale factors, and $e_0 = d_{31}/S_{011}$. In addition to the first and last

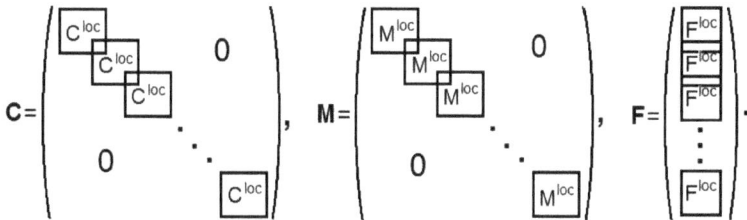

Fig. 3.2. Assembly of matrices and right parts.

elements of the global column vector of the right-hand side terms, it is necessary to add the terms $-\sigma/\Delta$, σ/Δ correspondingly, which are related to the normal mechanical stresses at the ends of the rod.

3.3.3. *Numerical implementation and analysis of solution*

As the main characteristic, we will study the amplitude–frequency characteristic of admittance Z, for which, we will find

$$Z = \frac{i\omega a l e_0^2}{hC_0} \sum_{k=1}^{N} \int_{-1}^{1} \left[\frac{e(\zeta)}{e_0} (N_1' \ N_2' \ N_3') \begin{pmatrix} u_{2k-1} \\ u_{2k} \\ u_{2k+1} \end{pmatrix} \frac{1}{l} + \frac{\ni(\zeta)C_0}{e_0^2} \Delta \right] d\zeta$$

(3.24)

If, instead of relations (3.12), we consider the case of uniform polarization $p_0 = p_{\text{sat}}$, $\varepsilon_{0\,\text{III}} = \varepsilon_{\text{sat}}$, then we get the well-known result of odd harmonics shown in Figure 3.3, where $\Omega_k = \frac{\pi}{2}(2k-1)$, $k = 1, 2, \ldots$.

Let us demonstrate some possibilities of the model, using the example of constructing the frequency response of the admittance for some variants of inhomogeneous polarization. Let us first consider piecewise inhomogeneous polarization along the length of the rod,

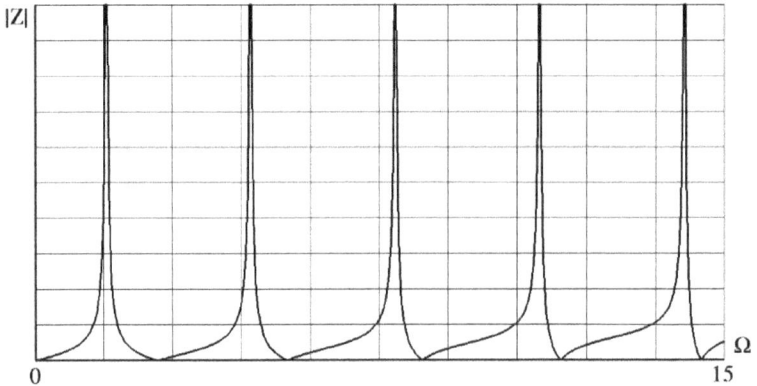

Fig. 3.3. Frequency response of admittance, uniform polarization.

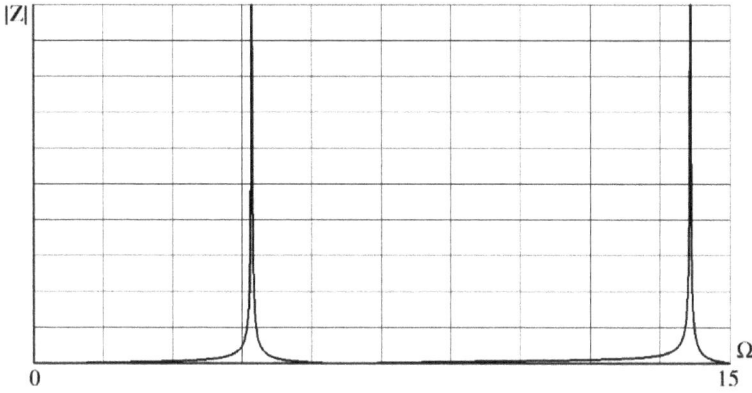

Fig. 3.4. Frequency response of admittance, piecewise constant polarization with three equal parts of its change in direction.

when its direction changes in three equal sections: (i) in the extreme sections, (ii) the polarization vector is directed upwards, and (iii) in the central part, it is directed downwards. The frequency response of the admittance of this case is shown in Figure 3.4.

If the frequency range is increased, then it can be noted that the frequency response will contain only 3, 9, 15, 21, 27, etc. frequencies. This behavior of the admittance frequency response suggests the idea of changing the residual polarization so that a response occurs only at certain frequencies. To this end, it was proposed to investigate the residual polarization and deformation in the form of certain harmonics, for example:

$$p_0 = p_{\text{sat}} \sum_{k=1}^{K} a_k \sin \frac{\pi k x}{l}, \quad \varepsilon_{0\,\text{III}} = \varepsilon_{\text{sat}} \sum_{k=1}^{K} a_k abs \left(\sin \frac{\pi k x}{l} \right), \quad (3.25)$$

where K is some finite number. We will study the cases of excitation of only one harmonic, for example, a harmonic with the number m. For this purpose, it is enough to put $a_m = 1$, and equate the remaining coefficients to zero. Figures 3.5–3.7 show the results for the first ($a_1 = 1$), third ($a_m = 3$) and fifth ($a_5 = 1$) harmonics.

Further calculations showed that it is possible to excite two or three modes, and even excite even-order harmonics. To do this, it is

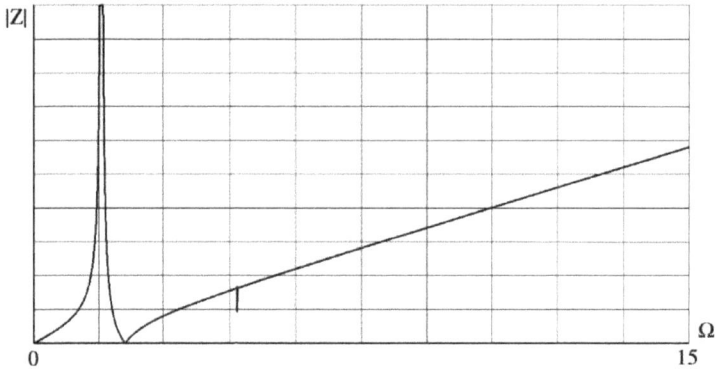

Fig. 3.5. Admittance frequency response, for polarization by the first harmonic of the sine $a_1 = 1$.

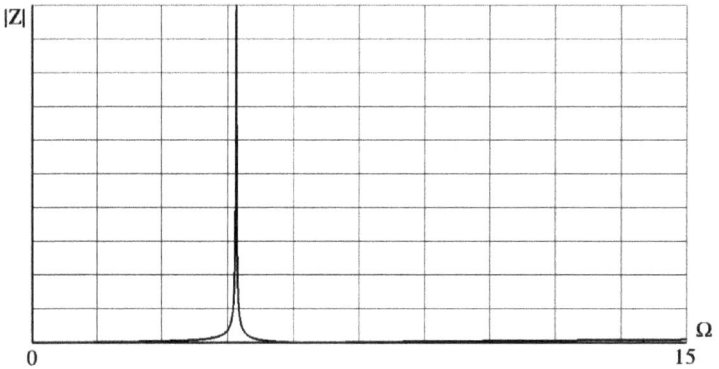

Fig. 3.6. Admittance frequency response, for polarization by the first harmonic of the sine $a_3 = 1$.

only required to choose the appropriate type of residual polarization and deformation. For example, Figure 3.8 shows the frequency response of the admittance for the second and fourth harmonics.

The performed calculations have shown that the inhomogeneous polarization significantly changes the picture of the admittance frequency response, allowing the corresponding law of inhomogeneity to achieve the suppression of certain and the generation of other harmonics.

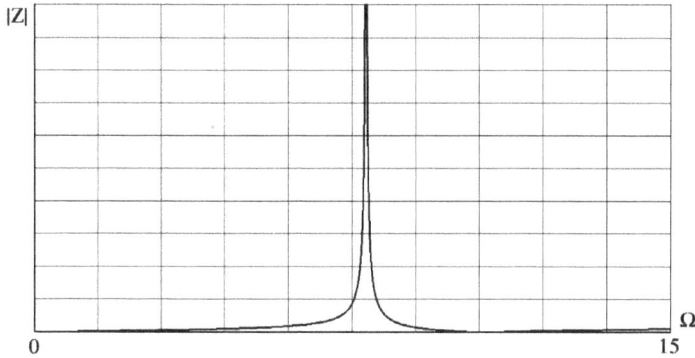

Fig. 3.7. Admittance frequency response, for polarization by the first harmonic of the sine $a_5 = 1$.

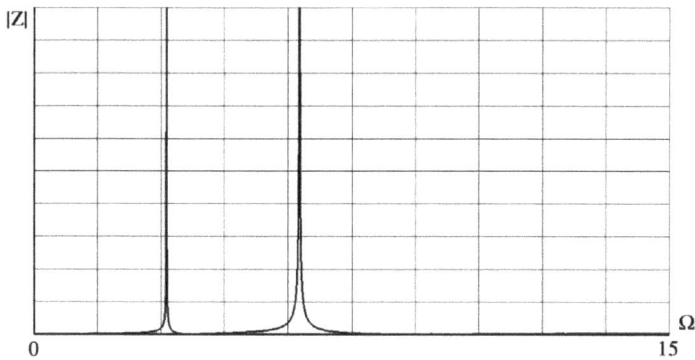

Fig. 3.8. Frequency response of admittance at polarization by the second and fourth harmonics of the sine $a_2 = 1$, $a_4 = 1$.

3.4. Acoustic Path with Source and Receiver of Vibrations

In this section, we consider a classical problem in a simplified formulation, in which we study a system consisting of an acoustic path with a source and a receiver of vibrations. Both the source and the receiver of vibrations are made of piezoceramic material. In the future, we will set the task to investigate the frequency nature of wave propagation in such a system, when the source and receiver of oscillations have non-uniform polarization and can generate selected oscillation modes.

However, to begin with, it will be interesting to obtain a solution to this problem but with piezoceramic transducers having a constant polarization. This section is devoted to this problem.

The study of wave processes, taking into account the mechanisms of attenuation, is an actual task that includes a heterogeneous physical medium. There are many problems on the excitation of waves in acoustic environments by piezoceramic transducers of various volumes and shapes, but there are few works that take account of the damping factors, associated with various mechanisms, which remove part of the energy from the system. Some of them include the phenomenon of hysteresis in a piezoceramic emitter and receiver, energy removal through the fasteners of the emitter and receiver, the viscosity of the fluid filling the acoustic waveguide and energy leakage through the elastic surface limiting the waveguide. In addition to this, we note that there are not so many works in which, along with the sources of waves, the receivers of these waves are considered, with a description of the information received, even under conditions of a harmonic oscillatory process. Initially, the use of vibration receivers was adapted in devices for ultrasonic thickness measurement and vibration and shock acceleration. However, in a vast area of wave processes, including waveguides, interest in such studies has increased only recently, which can be observed for non-classical and irregular regions, for example, for stratified acoustic waveguides [21, 324], waveguides with irregular walls [275], waveguides with impedance scattering [248] and waveguides with impedance walls [229].

In the present work, an attempt was made to investigate harmonic processes in an acoustic waveguide in the presence of a source and receiver of oscillations and the mechanisms of attenuation of propagating waves due to the removal of oscillation energy through the elastic walls of the waveguide.

The harmonic mode in the waveguide is generated by oscillating in the source, which is a thin piezoceramic transducer, the electrodes of which are supplied with differences of potentials by the harmonic law. After the waves pass through the acoustic part of the waveguide, the vibrations are picked up by a piezoceramic receiver. The case is considered when both the source and the receiver operate on thickness vibration modes. Thus, the system of equations for all the environments included in the system comprises the equations of electro-elasticity and the acoustic medium. To simplify the problem, both the emitter and the receiver are modeled by one-dimensional equations of the theory of electro-elasticity. However, in an acoustic waveguide, in order to take account of the impedance attenuation on the lateral surface, an axisymmetric problem for an acoustic fluid was considered. The conditions for conjugation of solutions at the interfaces between acoustic fluid and piezoceramic area are satisfied in the integral sense. The use of impedance boundary conditions on the side surface of the waveguide generates a spectral problem in which complex roots are found. For numerical simulation, only the piston mode was taken into account, which is why only one complex root was used. The boundary conditions for a piezoceramic sensor are supplemented by the equation of current flow through an external circuit, which includes an inductance, ohmic resistance and a capacitor, for which the complex conductivity of the external circuit was considered. In the obtained model, by controlling the parameters, it is possible to obtain special cases of the waveguide without taking into account the impedance attenuation and the influence of boundary conditions on the free wall of the sensor transducer, which made it possible to obtain quantitative estimates of the influence of these attenuation mechanisms in comparison with the ideal case.

3.4.1. *Design features and problem statement*

A mechanical system is considered, consisting of a piezoceramic emitter I, an acoustic waveguide II, a piezoceramic receiver III and a damping rubber pad IV, as shown in Figure 3.9.

On the electrodes of the converter I, volume Ω_1 from the voltage generator is supplied a difference of potentials by the harmonic law on time. The emerging thickness vibrations are transmitted to the acoustic waveguide II (volume Ω_2), which is a closed shell filled with liquid. The propagating perturbations, on the one hand, are captured by transducer III (volume Ω_3), and, on the other hand, are scattered through the waveguide boundary. A current arises in an external circuit with resistance Y, which connects with disturbances in the

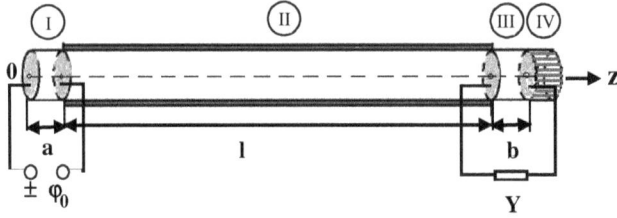

Fig. 3.9. General view of the acoustic path: I – piezoceramic transducer; II – acoustic waveguide; III – piezoceramic receiver; IV – damping rubber pad.

acoustic path. To reduce reflective effects, a foreign vibration damper in the form of a transition layer IV is glued to the right end of the receiver. It is required to determine the difference of the potentials on the electrodes of the right receiver and the generated current in the outer circuit.

To concretize the problem, we will make a set of assumptions and simplifications that simplify the mathematical solution but do not emasculate the physical essence of the phenomenon. First, we note that all the considered regions Ω_1, Ω_2, Ω_3 are cylinders, the lengths of which are equal to a, l, b, respectively, having a circular cross section of radius R_0. This allows us to consider the problem in an axisymmetric formulation. In the regions Ω_1 and Ω_3, the values a, b, R_0 are comparable with each other, and we have $R_0 \ll l$ in the region of the acoustic waveguide Ω_2. The transducers are considered as piezoceramic bodies, with the direction of preliminary polarization along the z-axis. The electrodes are considered thin enough; they do not affect the mechanical stresses. Any ideal or viscous liquid, such as water, can be used as the fluid in the acoustic waveguide. The surface that bounds the liquid is a solid elastic tube. The external electrical circuit on the receiver Ω_3 includes elements of resistance, capacitance and inductance, which allow us to consider its resistance Y as a complex value. Vibration damper IV is usually a rubber gasket.

In this section, we consider the problem for the acoustic path when the piezoceramic source and receiver have uniform polarization, with the polarization vector directed along the thickness of the transducers. To solve the problem, we use the equations of continuum mechanics: (i) it is an electro-elasticity model in the region Ω_1 and

Ω_3; (ii) it is an acoustic fluid model in the region Ω_2; (iii) vibration damper IV is well approximated by the viscoelastic Kelvin–Voigt model with the following constitutive relations:

$$\sigma = E\varepsilon + \gamma\dot{\varepsilon}. \tag{3.26}$$

The boundary conditions at the left end correspond to a rigid fixation, and at the right end, they satisfy the contact with an external viscoelastic environment. Moreover, it is necessary to use the conjugation conditions within the region for force and kinematic effects.

Next, we will consider the harmonic mode of oscillations:

$$\begin{aligned}
u_3(M,t) &= \hat{u}_3(M)e^{i\omega t}; & (\cdot)M \in S_1, S_3, \\
\varphi(M,t) &= \hat{\varphi}(M)e^{i\omega t}; & (\cdot)M \in S_1, S_3, \\
\psi(M,t) &= \hat{\psi}(M)e^{i\omega t}; & (\cdot)M \in S_2,
\end{aligned} \tag{3.27}$$

where u_3, φ are the axial displacement and the electric potential in Ω_1 and Ω_3, respectively; ψ is the acoustic potential in Ω_2; ω is the circular frequency; M are the points of the indicated areas. Variables with a "cap" are the amplitude components of the corresponding functions.

By using the aforementioned geometric features of our acoustic path in the regions Ω_1 and Ω_3, we write the electro-elasticity equations in the following one-dimensional formulation, which includes field equations, geometric relations and constitutive relations:

$$\left.\begin{aligned}
\frac{\partial \hat{\sigma}_{33}^{(1)}}{\partial z} &= -\rho\omega^2 u_3^{(1)}; \\[6pt]
\frac{\partial \hat{D}_3^{(1)}}{\partial z} &= 0; \\[6pt]
\hat{\varepsilon}_{33}^{(1)} &= \frac{\partial \hat{u}_3^{(1)}}{\partial z}, \\[6pt]
\hat{E}_3^{(1)} &= -\frac{\partial \hat{\varphi}}{\partial z}, \\[6pt]
\hat{\sigma}_{33}^{(1)} &= C_{33}^* \hat{\varepsilon}_{33}^{(1)} - e_{33}^* \hat{E}_3^{(1)}; \\[6pt]
\hat{D}_3^{(1)} &= e_{33}^* \hat{\varepsilon}_{33}^{(1)} + \vartheta_{33}^* \hat{E}_3^{(1)},
\end{aligned}\right\} M \in \Omega_1 \tag{3.28}$$

$$\left.\begin{aligned}
\frac{\partial \hat{\sigma}_{33}^{(3)}}{\partial z} &= -\rho \omega^2 u_3^{(3)}; \\[6pt]
\frac{\partial \hat{D}_3^{(3)}}{\partial z} &= 0; \\[6pt]
\hat{\varepsilon}_{33}^{(3)} &= \frac{\partial \hat{u}_3^{(3)}}{\partial z}, \\[6pt]
\hat{E}_3^{(3)} &= -\frac{\partial \hat{\varphi}}{\partial z}, \\[6pt]
\hat{\sigma}_{33}^{(3)} &= C_{33}^* \hat{\varepsilon}_{33}^{(3)} - e_{33}^* \hat{E}_3^{(3)}; \\[6pt]
\hat{D}_3^{(3)} &= e_{33}^* \hat{\varepsilon}_{33}^{(3)} + \vartheta_{33}^* \hat{E}_3^{(3)},
\end{aligned}\right\} \quad M \in \Omega_3 \qquad (3.29)$$

where ρ is the density of the piezoelectric ceramics of the source and receiver; $\hat{\sigma}_{33}$, $\hat{\varepsilon}_{33}$ are the stress and strain, respectively; $\hat{D}_3$, $\hat{E}_3$ are the electric displacement and electric field, respectively; C_{11}, C_{12}, C_{13}, C_{33} are elastic constants; e_{31}, e_{33} are piezoelectric coefficients; ϑ_{33} is the dielectric permeability of the ceramic material; and the new-made modules:

$$C_{33}^* = C_{33} - \frac{2C_{13}^2}{C_{11} + C_{12}}; \quad e_{33}^* = e_{33} - \frac{2C_{13}e_{31}}{C_{11} + C_{12}};$$

$$\vartheta_{33}^* = \vartheta_{33} - \frac{2e_{31}^2}{C_{11} + C_{12}} \qquad (3.30)$$

For an acoustic medium, we have

$$\frac{\partial^2 \hat{\psi}}{\partial r^2} + \frac{1}{r}\frac{\partial \psi}{\partial r} + \frac{\partial^2 \hat{\psi}}{\partial z^2} + \frac{\omega^2}{c^2\left(1 + i\omega\frac{\lambda_1 + 2\mu_1}{\rho_0 c^2}\right)}\hat{\psi} = 0, \quad M \in \Omega_2, \quad (3.31)$$

where ρ_0 is the density of the liquid; c is the speed of sound in the liquid; λ_1, μ_1 are viscosity coefficients. It should be noted that the numerical values of the dynamic viscosity coefficients for water are so small that they practically do not affect the attenuation process in a liquid. However, as they say "for the purity of the experiment" they are left in our model. The equations correspond to the piston oscillation mode in the domains Ω_1 and Ω_3, and Equation (3.31) is considered in a two-dimensional formulation in the Ω_2. This is necessary to take into account impedance attenuation through a cylindrical surface.

Let us consider in detail the boundary conditions and conjugation conditions. On the left side we have a rigidly fixed electrode:

$$\hat{u}_3^{(1)}\Big|_{z=0} = 0,$$
$$\hat{\varphi}^{(1)}\Big|_{z=0} = \varphi_0. \tag{3.32}$$

Condition at the right electrode of source is

$$\hat{\varphi}^{(1)}\Big|_{z=a} = -\varphi_0, \tag{3.33}$$

where $\pm\varphi_0$ are potentials at the electrodes of the source.

On the sidelong surface of the region Ω_1, the pressures and velocities of the contacting medium are equalized: $P_a = P_c$, $V_a = V_c$, which generate conditions of the impedance type

$$-\rho_0 i\omega\,\psi\big|_{r=R} - \frac{\partial\psi}{\partial r}\bigg|_{r=R}\,Z_c = 0. \tag{3.34}$$

At the right end of the receiver, we have

$$C_{33}^*\frac{\partial\hat{u}_3^{(3)}}{\partial z} + e_{33}^*\frac{\partial\hat{\varphi}^{(3)}}{\partial z}\bigg|_{z=a+l+b} = -\left(E\hat{\varepsilon}_{33}^{(3)} + \gamma i\omega\hat{\varepsilon}_{33}^{(3)}\right),$$
$$\hat{\varphi}^{(3)}\Big|_{z=a+l+b} = -\varphi_*. \tag{3.35}$$

Condition on the left electrode of the receiver is

$$\hat{\varphi}^{(3)}\Big|_{z=a+l} = \varphi_*, \tag{3.36}$$

where $\pm\varphi_*$ are potentials on the electrodes of the receiver, determined in the future; E, γ are the elastic and viscous coefficients of the Kelvin–Voigt model.

The conjugation conditions at the boundaries of the regions $\Omega_1 - \Omega_2$ and $\Omega_2 - \Omega_3$ are associated with the compatibility of the displacement and the continuity of stresses, as follows:

$$
i\omega \hat{u}_3^{(1)}\Big|_{z=a} = \frac{\partial \hat{\psi}}{\partial z}\Big|_{z=a},
$$

$$
C_{33}^* \frac{\partial \hat{u}_3^{(1)}}{\partial z} + e_{33}^* \frac{\partial \hat{\varphi}^{(1)}}{\partial z}\Big|_{z=a} = i\omega \rho_0 \, \hat{\psi}\Big|_{z=a},
$$

$$
i\omega \hat{u}_3^{(3)}\Big|_{z=a+l} = \frac{\partial \hat{\psi}}{\partial z}\Big|_{z=a+l},
$$

$$
C_{33}^* \frac{\partial \hat{u}_3^{(3)}}{\partial z} + e_{33}^* \frac{\partial \hat{\varphi}^{(3)}}{\partial z}\Big|_{z=a+l} = i\omega \rho_0 \, \hat{\psi}\Big|_{z=a+l}.
$$

(3.37)

However, relations (3.37) should be corrected, since the function on the left side does not depend (while on the right side it depends) on the radial coordinate. Let us consider these equations in more detail after we obtain the orthogonality conditions for the eigenfunctions.

Closing equation is the equation of Ohm's law for the section of the external circuit on the electrodes of the receiver:

$$
i\omega \int_S \hat{D}_3^{(3)} dS = \frac{2\varphi_*}{Y}.
$$

(3.38)

3.4.2. Solution, reduction to the solution of system of algebraic equations

The solution of equations in regions Ω_1, Ω_3 is

$$
\hat{u}_3^{(1)} = A \sin \frac{\Omega_1 z}{a} + B \cos \frac{\Omega_1 z}{a},
$$

$$
\varphi^{(1)} = \frac{e_{33}^*}{\vartheta_{33}^*} \left(A \sin \frac{\Omega_1 z}{a} + B \cos \frac{\Omega_1 z}{a} \right) + Cz + D, \quad z \in \Omega_1.
$$

(3.39)

$$
\hat{u}_3^{(3)} = G \sin \frac{\Omega_3 z}{b} + F \cos \frac{\Omega_3 z}{b},
$$

$$
\varphi^{(3)} = \frac{e_{33}^*}{\vartheta_{33}^*} \left(G \sin \frac{\Omega_3 z}{b} + F \cos \frac{\Omega_3 z}{b} \right) + Pz + Q, \quad z \in \Omega_3,
$$

(3.40)

where

$$\Omega_1^2 = \frac{\rho\omega^2 a^2}{C_{33}^* + \frac{(e_{33}^*)^2}{\vartheta_{33}*}}, \quad \Omega_3^2 = \frac{\rho\omega^2 b^2}{C_{33}^* + \frac{(e_{33}^*)^2}{\vartheta_{33}*}}. \tag{3.41}$$

The solution of the equation in the acoustic region Ω_2 is divided into two parts: $\hat{\psi} = R(r)Z(z)$. The first part generates the following spectral problem:

$$\frac{\partial^2 R}{\partial r^2} + \frac{1}{r}\frac{\partial R}{\partial r} + \left(\frac{\Omega_2^2}{l^2} - \kappa^2\right)R = 0,$$

$$\frac{\partial R}{\partial r} + \frac{i\omega\rho_0}{Z_c}R\bigg|_{r=R_0} = 0, \tag{3.42}$$

where

$$\Omega_2^2 = \frac{l^2\omega^2}{c^2\left(1 + i\omega\frac{\lambda_1 + 2\mu_1}{\rho_0 c^2}\right)}, \tag{3.43}$$

Z_c is the impedance of the environment.

The second part is reduced to the equation

$$\frac{\partial^2 Z}{\partial z^2} + \kappa^2 Z = 0. \tag{3.44}$$

The common solution must satisfy the interface conditions at the boundary with ceramic transducers.

To solve the spectral problem, it is required to determine the complex roots of the equation

$$J_1(a) - \frac{i\omega\rho_0 R_0}{aZ_c}J_0(a) = 0, \quad a = \sqrt{\frac{\Omega_2^2}{l^2} - \kappa^2} \tag{3.45}$$

and obtain orthogonality conditions for eigenfunctions.

3.4.3. *Numerical determination of roots*

Denote by $b = \frac{\omega\rho_0 R_0}{aZ_c}$ a dimensionless quantity, including the circular frequency, density, radius and acoustic impedance. Note that this value is proportional to the circular frequency. To study complex roots a_m, it is convenient to introduce the standard notation for a complex variable $z_m = x_m + iy_m$ and consider the equation

$$z J_1(z) - ib J_0(z) = 0 \tag{3.46}$$

for large and small values of this parameter. One can see that for small values of b, the roots will be close to the roots of the first-order Bessel function, and for large values, they will be close to the roots of the zero-order Bessel function. It is also easy to see that, along with the root z_m, the equation also contains the root $(-z_m)$. Using the left-hand side of the equation, we construct the following real function:

$$f(x, y) = \sqrt{(-z J_1(z) + ib J_0(z))(-\bar{z} J_1(\bar{z}) - ib J_0(\bar{z}))}, \tag{3.47}$$

where values with a bar at the top denote complex conjugate values, and those values z_m that provide the minimum of the constructed function will be the desired roots of the equation. The plot of this function for both large and small values of b is two surfaces over the first and third quadrants of the complex plane, as shown in Figure 3.10.

As follows from the figure, the constructed function has two branches of minimum values, each of which will be the desired root. As noted earlier, the real values of the roots are between the roots of the zero- and first-order Bessel functions, and, as the plot shows, the imaginary part takes small values. The small value of the imaginary part makes it possible to use asymptotic approximations for the approximate construction of roots. This approach was used by some authors in acoustic problems, for example, Ref. [229]. However, we will find the roots of the equation numerically using the variable step gradient descent method. Briefly, the algorithm is as follows. Let

$$[(z)] = -z J_1(z) + ib J_0(z), \quad [(\bar{z})] = -\bar{z} J_1(\bar{z}) - ib J_0(\bar{z}), \tag{3.48}$$

then $f(x, y) = \sqrt{[(z)][(\bar{z})]}$. To find the gradient of this function, we need to find its partial derivatives with respect to x and y. With this aim we find

$$\frac{\partial f(x, y)}{\partial x} = \frac{\partial f}{\partial z} + \frac{\partial f}{\partial \bar{z}}, \quad \frac{\partial f(x, y)}{\partial y} = \left(\frac{\partial f}{\partial z} - \frac{\partial f}{\partial \bar{z}} \right) i. \tag{3.49}$$

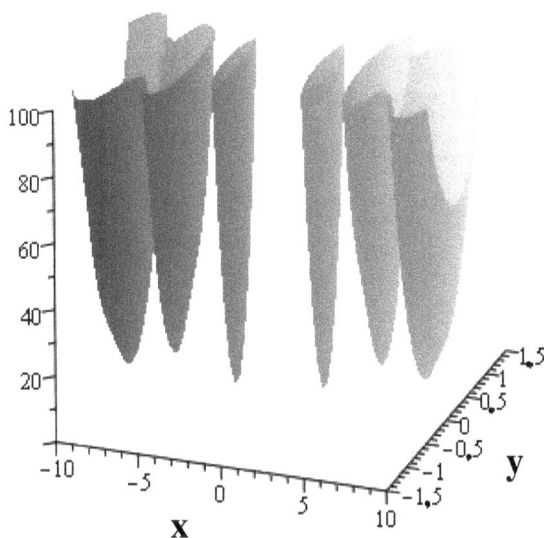

Fig. 3.10. Plot of the objective function for the study on the minimum value.

In its turn

$$\frac{\partial f}{\partial z} = \frac{\partial[(z)]}{\partial z} \frac{[(\bar{z})]}{2\sqrt{[(z)][(\bar{z})]}}, \quad \frac{\partial f}{\partial \bar{z}} = \frac{\partial[(\bar{z})]}{\partial \bar{z}} \frac{[(z)]}{2\sqrt{[(z)][(\bar{z})]}} \quad (3.50)$$

and

$$\frac{\partial[(z)]}{\partial z} = -z J_0(z) - ib J_1(z), \quad \frac{\partial[(\bar{z})]}{\partial \bar{z}} = -\bar{z} J_0(\bar{z}) + ib J_1(\bar{z}). \quad (3.51)$$

The numerical algorithm for finding the roots consists in constructing an iterative process with the choice of initial approximation (x_0, y_0) and setting numerical values for the parameters (λ_x, λ_y). According to Equations (3.49)–(3.51), the partial derivatives of the function for the values (x_n, y_n) are determined, and the values of the cosines are found as

$$\Delta_x = \frac{\frac{\partial f}{\partial x}}{\sqrt{\left(\frac{\partial f}{\partial x}\right)^2 + \left(\frac{\partial f}{\partial e}\right)^2}}, \quad \Delta_y = \frac{\frac{\partial f}{\partial y}}{\sqrt{\left(\frac{\partial f}{\partial x}\right)^2 + \left(\frac{\partial f}{\partial e}\right)^2}}. \quad (3.52)$$

The condition for each cosine from (3.32) is checked: does the sign change when passing to a new iteration? If the sign does not change, then the following approximations are determined:

$$x_{n+1} = x_n - \lambda_x \Delta_x, \quad y_{n+1} = y_n - \lambda_y \Delta_y, \quad n = 0, 1, \ldots. \quad (3.53)$$

If any of the cosines has changed sign, then the parameter λ with it decreases by a multiple with a denominator of 2 until there is no sign change. This approach avoids jumping from one edge of the "ravine" to the other. For example, for the parameter values

$$\omega = 10000 \, s^{-1}, \quad \rho_0 = 1 \cdot 10^3 \, kg/m^3, \quad R_0 = 0.01 \, m,$$
$$Z_c = 30 \, kg/(m^2 c)$$

after 100 iterations we get the values of the first three roots, presented in Table 3.1.

In this case, the parameter $b = 333.3$ is three orders of magnitude greater than unity. The roots are close to the roots of the zero-order Bessel function, and at the same time, the imaginary part has the third-order of smallness. Summing up, we can say that there is a countable set of roots of an equation with a positive real part, with a condensing point at infinity. The numerical method allows finding the first roots of an equation with a given accuracy. The initial values chosen as described above have practically no effect on the end solution. The asymptotic behavior of the roots with increasing number has not been studied, because oscillation modes close to the piston mode are being studied. However, within the framework of a holistic study of the problem, studies are carried out for the entire set of roots.

With this purpose, orthogonality properties of eigenfunctions were studied. Let $a_m = z_m$ and $a_k = z_k$ be two roots of the equation.

Table 3.1. The roots of the equation and their accuracy.

n	$z_m = x_m + iy_m$	$-z_m J_1(z_m) + ib J_0(z_m)$
1	$2.4048254 + 0.72144778 \cdot 10^{-3}i$	$-2.0 \cdot 10^{-9} + 0.82809188 \cdot 10^{-6}i$
2	$5.52007786 + 0.16560249 \cdot 10^{-2}i$	$-1.0 \cdot 10^{-9} - 0.21342083 \cdot 10^{-5}i$
3	$8.65372752 + 0.25961242 \cdot 10^{-2}i$	$+1.0 \cdot 10^{-8} + 0.44235853 \cdot 10^{-6}i$

The orthogonality conditions for the corresponding eigenfunctions can be obtained in the standard way from the spectral problem (3.42). These conditions will be used to satisfy the conjugation conditions (3.37). In general, these conditions take the form

$$R_0 \int_0^{R_0} \frac{r}{R_0} J_0\left(a_k \frac{r}{R_0}\right) J_0\left(a_m \frac{r}{R_0}\right) dr$$

$$= \begin{cases} 0, & k \neq m \\ \frac{1}{2}[J_0^2(a_k) + J_1^2(a_k)], & k = m. \end{cases} \tag{3.54}$$

Now the solution for the acoustic region can be written as follows:

$$\hat{\psi} = \sum_{m=1}^{\infty} J_0\left(\frac{r}{R_0} a_m\right) \left(M_m \sin \sqrt{\frac{\Omega_2^2}{l^2} - a_m^2} \frac{z}{l} + N_m \cos \sqrt{\frac{\Omega_2^2}{l^2} - a_m^2} \frac{z}{l}\right). \tag{3.55}$$

The obtained solutions contain $8 + 2 \cdot \infty + 1$ unknowns: 8 integration constants in regions Ω_1 and Ω_3, one unknown potential at the receiver electrodes, and $2 \cdot \infty$ integration constants in the acoustic zone of region Ω_2. Conversely, we have only 6 boundary conditions. To satisfy conjugation conditions (3.37), it is necessary to use orthogonality conditions (3.37), into which solution (3.55) must be substituted. In addition to these relations, it is necessary to add the equation for the current in the external circuit of the receiver (3.38). For the convenience of the further solution, we express the value of the unknown potential from the condition of the current of the external circuit as

$$\varphi_* = \frac{i\omega Y \pi R_0^2}{2} \left(e_{33}^* \frac{\partial u_3^{(3)}}{\partial z} - g_{33}^* \frac{\partial \varphi^{(3)}}{\partial z}\right)\Bigg|_{z=a+l+b} \tag{3.56}$$

and substitute it into the remaining boundary conditions and conjugation conditions. If only one root of the equation from the spectral problem is used, then

$$\hat{\psi} = J_0\left(\frac{a_1}{R}r\right) \left(M \sin \sqrt{\frac{\Omega_2^2}{l^2} - a_1^2} \frac{z}{l} + N \cos \sqrt{\frac{\Omega_2^2}{l^2} - a_1^2} \frac{z}{l}\right) \tag{3.57}$$

and conjugation conditions take the form

$$i\omega\hat{u}_3^{(1)}\Big|_{z=a} \int_0^{R_0} J_0\left(a_1\frac{r}{R_0}\right)rdr = \int_0^{R_0} \frac{\partial\hat{\psi}}{\partial z}\Big|_{z=a} J_0\left(a_1\frac{r}{R_0}\right)rdr,$$

$$\left(C_{33}^*\frac{\partial\hat{u}_3^{(1)}}{\partial z} + e_{33}^*\frac{\partial\hat{\varphi}^{(1)}}{\partial z}\right)\Big|_{z=a} \int_0^{R_0} J_0\left(a_1\frac{r}{R_0}\right)rdr$$

$$= i\omega\rho_0 \int_0^{R_0} \hat{\psi}\Big|_{z=a} J_0\left(a_1\frac{r}{R_0}\right)rdr,$$

$$i\omega\hat{u}_3^{(3)}\Big|_{z=a+l} \int_0^{R_0} J_0\left(a_1\frac{r}{R_0}\right)rdr = \int_0^{R_0} \frac{\partial\hat{\psi}}{\partial z}\Big|_{z=a+l} J_0\left(a_1\frac{r}{R_0}\right)rdr,$$

$$\left(C_{33}^*\frac{\partial\hat{u}_3^{(3)}}{\partial z} + e_{33}^*\frac{\partial\hat{\varphi}^{(3)}}{\partial z}\right)\Big|_{z=a+l} \int_0^{R_0} J_0\left(a_1\frac{r}{R_0}\right)rdr$$

$$= i\omega\rho_0 \int_0^{R_0} \hat{\psi}\Big|_{z=a+l} J_0\left(a_1\frac{r}{R_0}\right)rdr. \tag{3.58}$$

In this case, we obtain a system of 10 equations with complex coefficients:

$$a_{mn}x_n = b_n, \tag{3.59}$$

in which the vector of unknowns are constants of integration, the right-hand sides contain the known potential φ_0 at the source electrodes, and the matrix has the following structure:

$$A = \begin{pmatrix}
0 & a_{12} & 0 & 0 & 0 & 0 & 0 & 0 & 0 & 0 \\
0 & a_{22} & 0 & a_{24} & 0 & 0 & 0 & 0 & 0 & 0 \\
a_{31} & a_{32} & a_{33} & a_{34} & 0 & 0 & 0 & 0 & 0 & 0 \\
0 & 0 & 0 & 0 & 0 & 0 & a_{47} & a_{48} & a_{49} & 0 \\
0 & 0 & 0 & 0 & 0 & 0 & a_{57} & a_{58} & a_{59} & a_{5,10} \\
0 & 0 & 0 & 0 & 0 & 0 & a_{67} & a_{68} & a_{69} & a_{6,10} \\
a_{71} & a_{72} & 0 & 0 & a_{75} & a_{76} & 0 & 0 & 0 & 0 \\
a_{81} & a_{82} & a_{83} & 0 & a_{85} & a_{86} & 0 & 0 & 0 & 0 \\
0 & 0 & 0 & 0 & a_{95} & a_{96} & a_{97} & a_{98} & 0 & 0 \\
0 & 0 & 0 & 0 & a_{10,5} & a_{10,6} & a_{10,7} & a_{10,8} & a_{10,9} & 0
\end{pmatrix}.$$

The values of the matrix coefficients are not given owing to limited space.

3.4.4. *Analysis of the results of the problem with the acoustic path*

If we consider the case when the acoustic impedance on the side wall is $Z_c = \infty$, then, as already mentioned, the spectral equation is reduced to the form

$$J_1(a) = 0. \tag{3.60}$$

In this case, the roots of the equation of the spectral problem are real. Physically, this means that we are considering the low-frequency region of a one-dimensional problem for an acoustic waveguide with rigid walls.

But our task is to study the impedance attenuation on the side wall, so we will consider the acoustic impedance on the side wall: $Z_c \ll \infty$. The desired characteristic is the current in the external circuit of the receiver. Its expression after solving the system can be expressed in terms of one of the found integration constants, namely, in terms of P, which enters in relations (3.40) in the following form:

$$I = -i\omega\,\varphi_0 d^2\,\mathfrak{z}_{33}^* P. \tag{3.61}$$

Since this is a complex number, the current value will be characterized by the modulus of this value.

The $CTS-19$ (Russian abbreviation) ceramics were chosen for the source and receiver. It was considered that the acoustic waveguide is filled with water at $20°C$ with the following parameters: $\rho_0 = 1\cdot10^3\,(kg/m^3)$; $\lambda_1 = 2.421\cdot10^{-3}\,(Pa\cdot sec)$; $\mu_1 = 1.005\cdot10^{-3}\,(Pa\cdot sec)$.

Let the geometric dimensions of regions $\Omega_1 - \Omega_3$ take the following values: $a = 1.0\cdot10^{-2}\,(m)$; $l = 1.0\cdot10°\,(m)$; $b = 1.0\cdot10^{-2}\,(m)$; and the radius of the pipe $R_0 = 5.0\cdot10^{-3}\,(m)$. Complex resistance $Y = 1.0\cdot10^4 + 1.0\cdot10^2 i\,(Ohm)$. The impedance of the external environment $Z_c = 2.0\cdot10^7 \div 5.0\cdot10^7\,(kg/\sec m^2)$. Parameter values of the viscoelastic Kelvin–Voigt model $E = 2.5\cdot10^6 \div 3.4\cdot10^7\,(N/m^2)$; $\gamma = 1.4\cdot10^4 \div 4.0\cdot10^5\,Pa\cdot sec$.

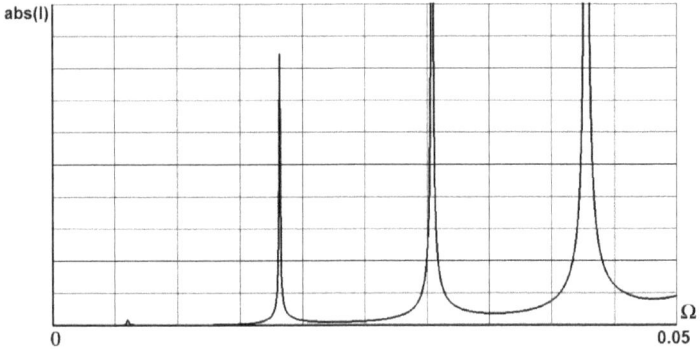

Fig. 3.11. Amplitude–frequency characteristic of the current module in the external circuit of the receiver.

Figure 3.11 shows the amplitude–frequency characteristics of the current module in the external circuit of the receiver in the frequency range $\Omega_1 \in [0.0, \ 0.5]$.

Note that the presence of a finite section of the acoustic path and the finite values of the thicknesses of the source and receiver of oscillations lead to re-reflection of waves in this system, and the appearance of a countable set of resonant bursts in the construction of the frequency response. The attenuation mechanisms in the form of an impedance at the side surface and a rubber gasket at the right end, modeled by the Kelvin–Voigt equations, lead to the fact that bursts of the current modulus at resonant frequencies are limited; the larger they are, the smaller the value of Z_c. The values of resonant frequencies are sensitive to these parameters and change noticeably with varying damping parameters.

3.5. The Problem for Acoustic Path with Source and Receiver of Vibrations in the Form of Non-uniformly Polarized Piezoceramic Transducers

We consider a mechanical system, consisting of piezoceramic transducer I, acoustic waveguide II, piezoceramic receiver III and damping gasket IV, shown in Figure 3.9. A distinctive feature of this problem from the above case is that the piezoceramic transducer and receiver are inhomogeneously polarized transducers along thickness. In this case, to solve the problem in regions I and III, it is necessary to use the finite element method, similarly to how it was done earlier, in the problem of longitudinal vibrations of a transversely polarized rod. Only in this case, there are two transducers of circular cross section with a remnant polarization vector along the z-axis. In contrast to the problem, described in Section 3.3.2, here the transducers perform longitudinal oscillations on the piezoelectric modulus e_{33}.

Leaving aside the generalized statement of the problem, we focus on finite element matrices. Local stiffness matrices for both region I and region III are easily found and have the block forms

$$\mathbf{C}^{(l)} = \begin{pmatrix} \mathbf{C}^{\mathrm{loc}} & \mathbf{e}^{\mathrm{loc}} \\ -\mathbf{e}^{\mathrm{loc}} & \mathbf{\mathfrak{z}}^{\mathrm{loc}} \end{pmatrix}, \quad \mathbf{M}^{(l)} = \begin{pmatrix} \mathbf{M}^{\mathrm{loc}} & 0 \\ 0 & 0 \end{pmatrix}, \tag{3.62}$$

where

$$\mathbf{C}^{\mathrm{loc}} = \frac{1}{\Delta^2} \int_{-1}^{1} C_{33}(\zeta) \begin{pmatrix} N_1' \\ N_2' \\ N_3' \end{pmatrix} (N_1' \ N_2' \ N_3') d\zeta,$$

$$\mathbf{e}^{\mathrm{loc}} = \int_{-1}^{1} e_{33}(\zeta) \begin{pmatrix} N_1' \\ N_2' \\ N_3' \end{pmatrix} (N_1' \ N_2' \ N_3') d\zeta,$$

$$\mathbf{\mathfrak{z}}^{\mathrm{loc}} = \int_{-1}^{1} \mathfrak{z}_{33}(\zeta) \begin{pmatrix} N_1' \\ N_2' \\ N_3' \end{pmatrix} (N_1' \ N_2' \ N_3') d\zeta,$$

$$M^{\mathrm{loc}} = \int_{-1}^{1} \begin{pmatrix} N_1 \\ N_2 \\ N_3 \end{pmatrix} (N_1 \ N_2 \ N_3) d\zeta.$$

$$(3.63)$$

Form functions were introduced earlier in Section 3.3.2.

The solution for the acoustic part is the same exactly as described in Section 3.4. The main difficulty lies in the formation of common stiffness and mass matrices. Obviously, these matrices will have a block structure.

If the length of a segment equal to the thickness of the transducers divides into k finite elements, then there will be $2k + 1$ nodes on this element. Now we introduce the following auxiliary matrices and vectors with the dimension indicated in brackets. These include stiffness matrices, associated with elastic, piezoelectric and dielectric constants and a matrix with acoustic zone coefficients as follows:

$$\mathbf{C}_{(2k+1)x(2k+1)}, \quad \mathbf{e}_{(2k+1)x(2k+1)}, \quad \mathbf{\ni}_{(2k+1)x(2k+1)}, \quad \mathbf{H}_{(2x2)}, \quad \mathbf{o}_{(2x2)}. \tag{3.64}$$

Next, you need to enter zero matrices of the indicated sizes:

$$\mathbf{O}_{(2k+1)x(2k+1)}, \quad \mathbf{O2}_{(2k+1)x(2k+1)} \tag{3.65}$$

and vectors, consisting of unknown nodal displacements, nodal potentials in regions I and III, the vector of unknowns in the acoustic region, and the right-hand sides of the corresponding equations:

$$\begin{aligned} &\mathbf{u}^{I}_{(2k+1)}, \quad \varphi^{I}_{(2k+1)}, \quad \mathbf{u}^{III}_{(2k+1)}, \quad \varphi^{III}_{(2k+1)}, \quad \mathbf{a}_{(2)}, \\ &\mathbf{b}^{I}_{u(2k+1)}, \quad \mathbf{b}^{I}_{\varphi(2k+1)}, \quad \mathbf{b}^{III}_{u(2k+1)}, \quad \mathbf{b}^{III}_{\varphi(2k+1)}, \quad \mathbf{b}_{(2)}. \end{aligned} \tag{3.66}$$

After carrying out the operation of assembling common matrices, the solution of the problem is reduced to solving the system of linear equations

$$(\mathbf{A} - \Omega^2 \mathbf{B}) \cdot \mathbf{U} = \mathbf{F}, \tag{3.67}$$

where

$$\mathbf{A} = \begin{pmatrix} \mathbf{C} & \mathbf{e} & \mathbf{O}_2^T & \mathbf{O} & \mathbf{O} \\ -\mathbf{e} & \mathbf{\ni} & \mathbf{O}_2^T & \mathbf{O} & \mathbf{O} \\ \mathbf{O_2} & \mathbf{O_2} & \mathbf{H} & \mathbf{O_2} & \mathbf{O_2} \\ \mathbf{O} & \mathbf{O} & \mathbf{O}_2^T & \mathbf{C} & \mathbf{e} \\ \mathbf{O} & \mathbf{O} & \mathbf{O}_2^T & -\mathbf{e} & \mathbf{\ni} \end{pmatrix},$$

$$\mathbf{B} = \begin{pmatrix} \mathbf{M} & \mathbf{O} & \mathbf{O}_2^T & \mathbf{O} & \mathbf{O} \\ \mathbf{O} & \mathbf{O} & \mathbf{O}_2^T & \mathbf{O} & \mathbf{O} \\ \mathbf{O_2} & \mathbf{O_2} & \mathbf{o} & \mathbf{O_2} & \mathbf{O_2} \\ \mathbf{O} & \mathbf{O} & \mathbf{O}_2^T & \mathbf{M} & \mathbf{O} \\ \mathbf{O} & \mathbf{O} & \mathbf{O}_2^T & \mathbf{O} & \mathbf{O} \end{pmatrix} \tag{3.68}$$

$$\mathbf{U} = \begin{pmatrix} \mathbf{u}^I & \varphi^I & \mathbf{a} & \mathbf{u}^{III} & \varphi^{III} \end{pmatrix}^T, \quad \mathbf{F} = \begin{pmatrix} \mathbf{b}^I_u & \mathbf{b}^I_\varphi & \mathbf{b} & \mathbf{b}^{III}_u & \mathbf{b}^{III}_\varphi \end{pmatrix}^T.$$

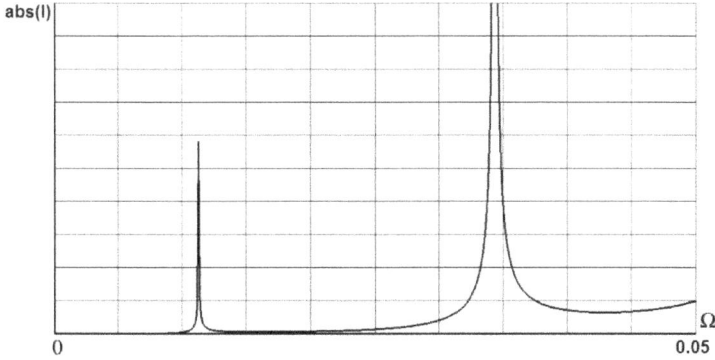

Fig. 3.12. Amplitude–frequency characteristic of the current module in the external circuit of the receiver with non-uniformly polarized transducers.

However, account of the boundary conditions and conjugation conditions requires the reorganization of the main matrix and right-hand sides. Taking into account that the left edge of the source is rigidly fixed, it is necessary to put the elements $a_{1,1} = 1$, $f_1 = 0$, and equate all other elements of the first row to zero. To take into account the potential at the left and right ends of the source transducer, it is necessary to set $a_{2k+2,2k+2} = 1$, $f_{2k+2} = \varphi_0$, $a_{4k+2,4k+2} = 1$, $f_{4k+2} = -\varphi_0$, and all other elements of these lines to equate to zero. Similar operations are carried out for other elements of the matrix, so the elements of the matrix $\mathbf{H}$ will change.

Having chosen the law of longitudinal preliminary polarization of the source and receiver converters in the form of one first harmonic, we calculated the frequency response of the current of the external circuit of the receiver. The results are shown in Figure 3.12.

We will leave a more detailed analysis of the influence of all geometric and physical parameters on the frequency response of the current until the next research.

3.6. Conclusions

The main positions for determining the physical properties of ferroelectric ceramics with non-uniform polarization are considered. It is shown that even in the one-dimensional case, a non-uniformly polarized piezoceramic rod can change its physical dynamic characteristics compared to a rod with a uniform polarization.

The problem of excitation and reception of waves in an acoustic waveguide by piezoceramic transducers with uniform polarization is solved. The acoustic waveguide was considered under conditions of energy loss through the side wall via impedance attenuation. The presence of impedance boundary conditions on the lateral surface of the waveguide generates a spectral problem in which complex roots were determined. A gradient descent algorithm was used to determine them. The low-frequency spectrum of oscillations allows us to consider the piston mode, for which only one root of the equation is used. The boundary conditions for the receiver are supplemented by the equation of current flow through an external circuit, including inductance, ohmic resistance and a capacitor, by choosing the complex conductivity of the external circuit. The current in the external circuit of the receiving transducer is investigated. The conjugation conditions are satisfied in the integral sense using the orthogonality conditions for the eigenfunctions. Further, a similar problem is investigated, when piezoceramic transducers have non-uniform polarization. The numerical experiments have shown that inhomogeneous polarization can reduce the number of resonant frequencies.

The results of the work can be used for medical purposes to study blood flow in vessels with elastic walls. They are also of interest when modeling a non-stationary solution to a problem.

Chapter 4

Piezoelectric Energy Harvesting: Main Directions of Development and Achievements

4.1. Introduction

In the 21st century, due to the accelerated development of what has come to be known as "green energy", the scientific direction of energy harvesting was finally formed. It is based on achievements in the development of new materials and composites, appropriate transducers and methods of transformation of environmental energy into electricity. An important element of these studies is the development of efficient electrical circuits with minimal energy losses during energy transformation and transfer between separate devices and networks. This chapter discusses some main directions of piezoelectric energy harvesting and important research results, obtained in the last decade. Theoretical, experimental and computer modeling approaches and methods are discussed in necessary detail.

Piezoelectric devices based on direct piezoelectric effect (transformation of mechanical energy into electric energy) and inverse piezoelectric effect (transformation of electric energy into mechanical energy) have found broad applications in the processes of energy harvesting. Piezoelectric generators (PEGs) or harvesters of different types are the main representatives of devices with direct piezoelectric effect and piezoelectric actuators of various kinds are the main representatives of devices with inverse piezoelectric effect. Their joint operation in the framework of the common systems can increase the efficiency of energy harvesting.

This chapter is organized as follows. Section 4.2 discusses state-of-the-art of piezoelectric generators research and development. In Section 4.3, the results and achievements in experimental investigations and modeling of cantilever-type PEGs obtained in the last decade are presented. Special attention is devoted to generators with proof masses and generators based on porous piezoceramics. Section 4.4 considers the problems of nonlinearity and expanding the frequency broadband to increase the efficiency of energy harvesters. Specific discussion touches upon monostable, bistable, tristable and multistable harvesters, taking into account obtained results and future problems in these directions. Section 4.5 considers PEGs with constructive L-shaped elements, presenting results of finite element modeling (FEM) of axial-type piezogenerators. The results for stack-type PEGs are examined in Section 4.6. Section 4.7 discusses some medical applications, based on piezoelectric transducers, namely, in drug delivery, blood sampling and ophthalmology. Finally, Section 4.8 presents some conclusions to Chapter 4.

4.2. State-of-the-Art of Piezoelectric Generators Research and Development

In the environment, vibrations are present in all types of constructions, both natural and artificial. The transformation of mechanical vibrations into electrical energy is an important issue. One of the types of generators (or harvesters) are piezoelectric converters of mechanical energy into electrical energy, called the piezoelectric generators (PEGs). The basic information about PEGs, as well as the problems arising at the various stages of design and research of energy harvesting devices, have been presented in reviews [11,57,98], as well as in fundamental monographs [136,358].

Today, the use of PEGs is actual for several classes of devices, such as: (i) small-sized wireless electronics with a long service life; (ii) low-power built-in and wireless communication devices (for example, phones, walkie-talkies); (iii) household electrical appliances and electronics (for example, an electronic clock); (iv) sensors and tracking systems operating in certain climatic zones or places inaccessible to humans and (v) various transducers for lighting and alarm systems [129, 354, 417].

Depending on different applications, various types of PEGs have been created, in which the direct piezoelectric effect is used when excited in the sensitive element as rule by longitudinal (d_{33}) [357, 374, 449, 493] or bending (d_{31}) [6, 8, 384] oscillations.

The problem of evaluating the efficiency of the used and developed piezoelectric materials has been considered in some works [135, 197]. These authors presented methods for measuring output characteristics of devices, as well as the parameters of piezomaterials with new electro-physical and mechanical properties.

Currently, much attention is being paid to the research and development of renewable energy sources, which use piezoceramic and thin-film piezoelectric elements (PEs) that convert environmental energy (vibrations, wind, solar energy, heat, radiation, aero fluxes and fluid flows, etc.) into electrical energy with its subsequent accumulation and transmission to the receiver. An analysis of the requirements for energy sources (generators, actuators and environmental energy converters) showed that they significantly depend on used types of devices. Nevertheless, most piezoelectric energy harvesting systems for low-power consumers of electricity, received from the environment, consist of functionally homogeneous components, the structure of which is represented in Figure 4.1.

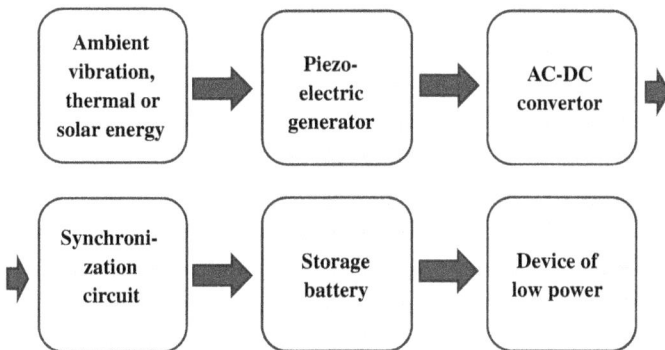

Fig. 4.1. Block-scheme of piezoelectric energy harvesting system of the environment.

To date, the problem of creating PEGs for various purposes has not been solved completely due to the low energy efficiency of PEGs and the low output power of the developed devices. Among the various problems encountered in the development of PEGs, the most significant are related to: (i) the choice of energy-efficient piezoceramic materials and composites, (ii) the development of electrical circuits for the accumulation of electrical energy with minimal charge leakage, (iii) the search for geometric configurations and assembly technologies of the PEG sensing element that provide maximum output power.

In relation to the first research direction, analytical and computer simulation (including finite element) approaches have already been intensively developed. In particular, analytical and FE modeling of porous piezoelectric ceramics has been carried out in [385], non-uniform polarization of multilayered piezoelectric transducers has been studied in [380], and structures, consisting of transversely polarized and longitudinally polarized parts, have been investigated in [375].

The research approaches, used in solving these problems, differ significantly depending on the field and specifics of the PEG application. This chapter will discuss in more detail the relevant experimental, analytical and model solutions with corresponding results, related to the above second and third research directions, which have been obtained over the past decade for cantilever-type, axial-type and stack-type PEGs.

4.3. Cantilever-type Piezoelectric Generators

4.3.1. *Experimental and model studies of cantilever-type PEGs*

The problem of estimating the energy efficiency of a cantilever-type PEG was previously considered in [6, 8, 136]. It has been shown that the output power of a PEG depends not only on the electrical characteristics of the piezoceramic materials in the sensitive PEG elements, but also on the measurement technique of their output characteristics as well as on the parameters of the electrical circuit [135].

The authors of the book have participated in the development of an experimental approach for estimating the output parameters of a cantilever piezoelectric power generator with active structures in the form of a bimorph [8, 9, 65, 319, 383].

The whole lot of cantilever setups under harmonic excitation, with different proof mass locations and electric load resistances, have been studied in experiments:

(i) *Cantilever PEG with proof mass*: The developed measurement system has allowed studying the output parameters (voltage and power) of PEG under vibrational excitation by using certificate Russian software [320, 321], generating a signal by scanning and recording it on a computer.

(ii) *Double-cantilever PEG with proof masses*: The specific output electric power of this PEG model was equal to 69.2 mW/cm^3 [9], which was three orders of magnitude higher in comparison with the cantilever PEG studied in [8].

Resonance piezoelectric harvesters have restrictions owing to their narrow bandwidth (typically equal to few hertz), so they can operate efficiently only at or near their natural frequency (defined by the device geometry and materials).

There are some passive frequency tuning devices, such as a moveable clamp that changes the beam length, or nonlinear, bistable structures with destabilizing axial loads (see, for example, Figure 4.2).

At the same time, active tuning harvesters were proposed in [93, 216, 461]. In particular, an actuated harvester using piezoelectric polycrystalline (PZT) components both for harvesting and actuation has been developed in [93]. It consisted of a two-beam device with three arms. In such a harvester, the central and larger arm housed the piezoelectric actuator, while two piezoelectric harvesters were located in the lateral tight arms. The authors reported that this device is capable of harvesting up to 90 μW at an acceleration amplitude of 0.6 g. The natural frequency was tuned into the

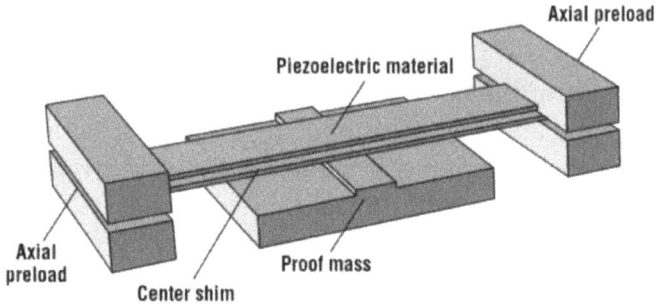

Fig. 4.2. Cantilever with axial load, presenting a passive frequency tuning device.

range from 150 to 215 Hz with actuator voltages between -30 and $+45$ V.

Taken as a whole, the piezoelectric actuation has been adopted to solve these problems due to its accuracy, time of response, low power consumption and its suitability for various uses. Therefore, important applications of the piezoelectric actuators are connected with three main stages associated with piezoelectric energy harvesting [419]: (i) mechanical–mechanical energy transfer, including mechanical stability of the piezoelectric device under big stresses, and mechanical impedance matching, (ii) mechanical–electrical energy transduction, related with the electromechanical coupling factor in the composite harvester structure, and (iii) electrical–electrical energy transfer, including electrical impedance matching, such as an AC/DC converter to accumulate the energy into a rechargeable battery.

There are several ways of modeling PEG: (i) mathematical modeling with lumped parameters [493], (ii) mathematical modeling with distributed parameters [374, 394, 449] and (iii) finite element (FE) modeling [357, 384]. The authors have performed the cantilever-type PEG modeling for some structures with various physical conditions of operating the PEG and its various geometries.

In several works, the theoretical and numerical approaches to the study of PEGs have been presented [58, 63, 327, 376, 383, 384]. The efficiency of these approaches was shown by close results in comparison with experiments. The modeling of the cantilever-type

PEG with symmetrical and asymmetrical locations of proof mass was based on the linear theory of elasticity and electrodynamics, taking into account the dissipation of energy, as well as the equations of motion in the acoustic approximation [384]. The built FE models were numerically realized in the ANSYS software.

After the modal analysis was performed, two PEG models with different conditions of the proof mass attachment and mechanical excitation were considered. The first model performed small oscillations in the moved coordinate system, associated with the surface, which was located at the left side of the plate. In the second model, the PEG base could move freely only in the vertical direction, while the horizontal direction was fixed. The vibrations were excited by the external force.

Numerical optimization of a cantilever-type PEG with incomplete covering of substrate by PEs has been carried out by using the developed analytical model [319, 376]. It has been based on the Kantorovich method [374], confirmed by FE modeling and a series of experiments [358]. Finite element analysis of the improved model of cantilever PEG has been performed, too [383]. In [376], the adequacy of using the straight normal hypothesis in the applied theory for the calculation of the characteristics of cantilever-type PEG with incomplete piezo-element coating of the substrate was stated. The calculations applied a numerical optimization procedure for PEG, taking into account critical stresses.

The optimization results showed that there is a frequency range where the first resonance frequencies of the PEG with and without the proof mass coincide, but the PEG model in absence of the proof mass could not operate in this range owing to the fact that the threshold of critical stresses had been exceeded.

In [383], the linear theory of elasticity and electrodynamics, based on energy dissipation, were applied together with the equations of motion of liquids and gases in the acoustic approximation for modeling a double-cantilever PEG with symmetrically located proof masses. The FE analysis showed that for 1 and 2 vibration modes, the maximum output power of 669 μW and 720 μW, respectively, was achieved at an electric load resistance of $R = 50\,k\Omega$.

Finally, a model of artificial grass (a cantilevered structure), designed to convert wind energy to electrical energy, was performed in [142]. Modal analysis was carried out for three cases of bimorph cantilever beams and natural frequencies, namely (i) without proof masses, (ii) with steel proof masses and (iii) with duralumin proof masses. The results obtained showed that the natural frequency of the cantilever energy harvester with the duralumin proof masses was equal to 29.88 Hz. The peak voltage of 44.99 V took place at a wind pressure of 1 Pa. The artificial grass harvested energy of about 0.0123 W and total output power was equal to 12.95 W.

4.3.2. *Cantilever-type PEG, based on porous piezoceramic, with proof mass*

The method of effective moduli in combination with the finite element solution of homogenization problems and modeling of representative volumes makes it possible to fully take into account the internal structure of the piezocomposite, types of its connectivity, as well as the shapes and sizes of inclusions or pores [166, 176].

In [381], FE modeling of oscillations of a cantilever-type PEG with an active pinch, which had piezoelements with effective properties of piezoelectric ceramics of a certain porosity, has been carried out.

The linear equations of the theory of elasticity and electroelasticity were used, taking into account the energy dissipation, adopted in the ANSYS software [288], as well as the motion equations of liquid and gaseous media in the acoustic approximation. For a piezoelectric medium, the mathematical formulation of the problem was described in the form of the following equations:

$$\rho \ddot{u}_i + \alpha \rho \dot{u}_i - \sigma_{ij,j} = f_i; \quad D_{i,i} = 0,$$
$$\sigma_{ij} = c_{ijkl}(\varepsilon_{kl} + \beta \dot{\varepsilon}_{kl}) - e_{ijk} E_k,$$
$$D_i + \varsigma_d \dot{D}_i = e_{ikl}(\varepsilon_{kl} + \varsigma_d \dot{\varepsilon}_{kl}) + \ni_{ik} E_k, \quad (4.1)$$
$$\varepsilon_{kl} = (u_{k,l} + u_{l,k})/2,$$
$$E_k = \varphi_k,$$

where ρ is the density of the material; u_i are the components of the vector-function of displacements; σ_{ij} are the components of the tensor of mechanical stresses; f_i are the components of the density vector of mass forces; D_i are the components of the electric induction vector; c_{ijkl} are the components of the fourth rank tensor of elastic moduli; e_{ikl} are the components of the third rank tensor of piezomodules; ε_{kl} are the strain tensor components; E_k are the components of the vector of the electric field strength; φ_k is the electric potential; $\mathfrak{z}_{ik}$ are the components of the second rank tensor of dielectric permittivity; $\alpha, \beta, \varsigma_d$ are non-negative damping factors (in ANSYS, $\varsigma_d = 0$).

For an elastic medium, the mathematical formulation of the problem was described in the following form:

$$\rho\ddot{u}_i + \alpha\rho\dot{u}_i - \sigma_{ij,j} = f_i,$$

$$\sigma_{ij} = c_{ijkl}(\varepsilon_{kl} + \beta\dot{\varepsilon}_{kl}), \qquad (4.2)$$

$$\varepsilon_{kl} = (u_{k,l} + u_{l,k})/2.$$

To solve dynamic problems of acoustoelectroelasticity, the FE model with the classical Lagrangian formulation was used.

The finite element formulation of the problem can be presented based on the example in [288]. In the transition from the continuous formulation to the FE one, a coordinated discretization of geometric regions was performed, that is, a partitioning into finite elements (triangulation) with a certain set of geometric points being nodes, as a result of which (4.1), (4.2) was represented as

$$u(x,t) = N_u^T(x)U(t); \quad \varphi(x,t) = N_\varphi^T(x)\Phi(t),$$

$$\psi(x,t) = N_\psi^T(x)\Psi(t), \qquad (4.3)$$

where x is the vector of spatial coordinates; t is the time; N_u^T is the matrix of shape functions for the displacement field: $u = (u_1, u_2, u_3)$; N_φ^T, N_ψ^T are the row vectors of the shape functions for the fields of the electric potential φ and the velocity potential in the acoustic medium ψ; $U(t)$, $\Phi(t)$, $\Psi(t)$ are the global vectors of the corresponding nodal degrees of freedom.

An approximation of FE model (4.3) for generalized statements of dynamic problems (4.1), (4.2), including boundary conditions, was reduced to a system of ordinary differential equations with respect to the nodal unknowns $a = [U, \Phi, \Psi]^T$, as follows:

$$M\ddot{a} + C\dot{a} + Ka = F, \qquad (4.4)$$

where the global matrices M, C, K for elastic medium are, respectively, the matrices of mass, damping and stiffness.

The properties of a piezoceramic material, used for modeling, are presented in [289, 290]. An analog of ceramics with such properties is the piezoceramic PZT-4. The electrical scheme of the PEG connection with an active load is shown in Figure 4.3.

The structure and FE-scheme of PEG is shown in Figure 4.4. Figures 4.5 and 4.6 show the results for the output voltage U in dependence on the frequency of harmonic excitation of the device. They correspond to the voltage values on the bimorph plates, disposed on the PEG cantilever substrate, and on the PEG piezoelectric cylinders, respectively.

The result of the performed modal and harmonic analysis of the computations for six vibration modes demonstrated that the most effective modes were 1 and 3. They had bending oscillation modes in the plane of the least stiffness. The analysis showed that by comparing the output characteristics of the ceramics with

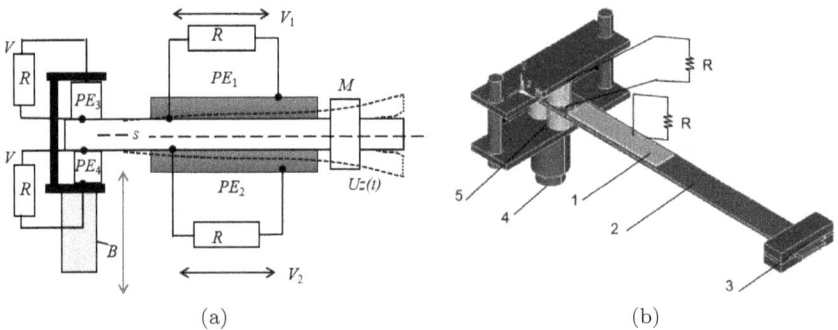

Fig. 4.3. (a) Electric scheme of PEG under active load; (b) location of electrical load on model: (1) PE; (2) substrate; (3) proof mass; (4) place of PEG fixing (B is the movable pinch); (5) piezoelectric cylinders.

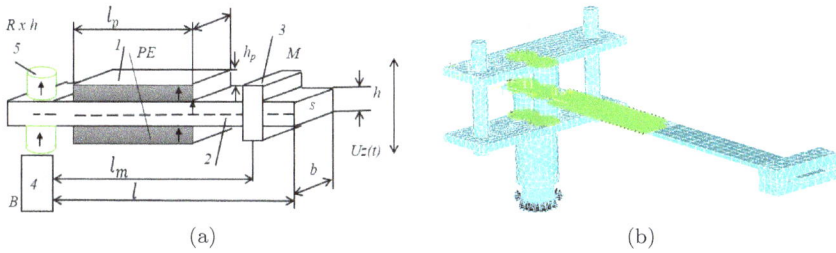

Fig. 4.4. (a) Structure and (b) FE-scheme of PEG with proof mass: (1) PE; (2) substrate; (3) proof mass; (4) place of PEG fixing (*B* is the movable pinch); (5) piezoelectric cylinders.

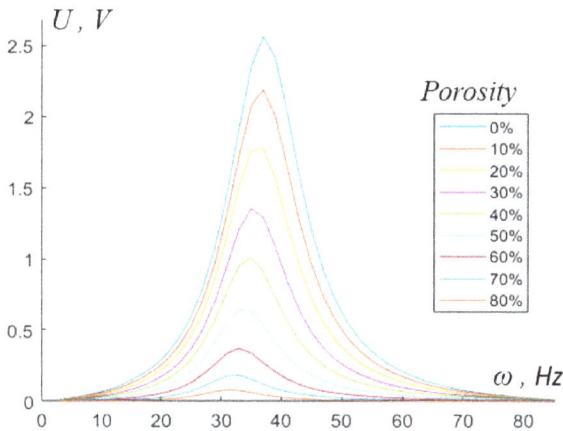

Fig. 4.5. Output voltage on the bimorph plates, located on the PEG cantilever substrate in dependence on the frequency of harmonic excitation; frequency range from 0 to 80 Hz.

a porosity of 0% and 80%, the output voltage of the structure decreased by 31 times. At the same time, the output power at an electric load of 1000 Ω decreased by more than 950 times. Cylindrical PEs at the PEG pinch were more effective for comparison. By comparing the output characteristics of these ceramic structures, in the constructions with a porosity of 0% and 80%, the output voltage changed by 6.8 times, while the output power changed by 47 times, respectively. More complete results for this cantilever-type PEG will be presented in Chapter 5.

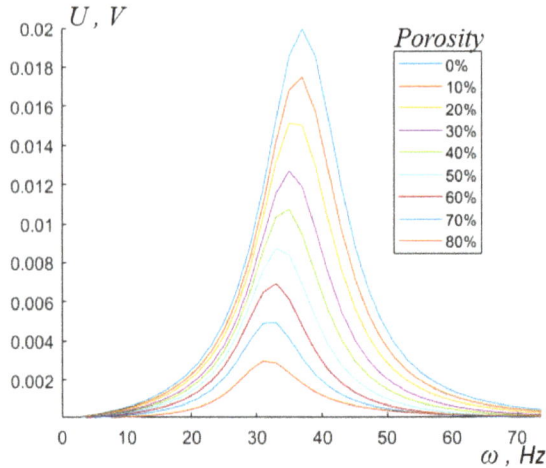

Fig. 4.6. Output voltage on the PEG piezoelectric cylinders in dependence on the frequency of harmonic excitation; frequency range from 0 to 70 Hz.

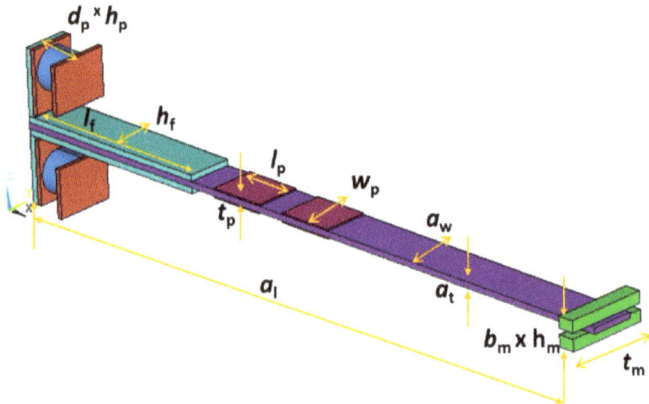

Fig. 4.7. Schematic diagram of piezoelectric energy harvester.

Another axial-type PEG with different values of porosities (changed from 0% to 80% through thickness or along the length of the duralumin beam) of piezoelectric ceramics was designed and studied in [138]. The axial-type energy harvester consisted of bimorph (d_{31}) and cylinder (d_{33}) piezoelectric patches with base excitation (see Figure 4.7). Based on the three-dimensional finite element modeling, the effects of various porosities, proof mass locations and different applied accelerations were used to determine the output voltage and output power generation. The maximum values of the voltage and power were obtained equal to 2.25 V and 5.1 μW, respectively.

4.4. Nonlinearity and Broadband Energy Harvesters

The proposed harvesters of oscillation energy are often based on linear mechanical principles. Such devices give appreciable response amplitude only if the dominant ambient vibration frequency is close to the resonance frequency of the harvester. Consequently, an output power decreases rapidly away from the resonant frequency in energy harvesters, operating on the basis of resonance. Therefore, to achieve maximum conversion efficiency, the dominant ambient vibration frequency should be known before the design process.

For a broadband or time-varying ambient oscillation spectrum, only a small fraction of the available ambient vibration energy can be extracted by such transducers. To overcome this restriction, a number of strategies have been developed to increase the efficiency of energy harvesters [411, 503], including the use of tunable resonators, multifrequency arrays and nonlinear generators.

Exploiting nonlinearity has become an alternative solution for broadband energy harvesting. First, two main approaches were developed: (i) designing a harvesting system with the aim to harden frequency response at periodic loading [28,249,332] and (ii) designing a harvesting system with a double-well potential (bistable system), which can jump between the wells at periodic or stochastic loading [74,97,189]. Subsequently, these approaches have been expanded and different structures were designed for piezoelectric energy harvesters, possessing monostable [101,210,219,291,454], bi-stable [75,130,447, 490], tri-stable [218,474,500], quad-stable [446,502] and even quin-stable [445] characteristics.

4.4.1. *Monostable harvesters*

One of the requirements for energy harvesting is to harvest reasonable amount of energy at low excitation frequency (for example, at the vibration of long span bridges or tall buildings). In [111], an inverted

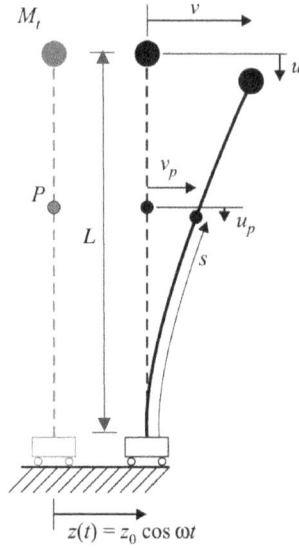

Fig. 4.8. Schematic representation of an inverted beam harvester: M_t is the proof mass at the tip of the elastic beam with length L under harmonic base excitation: $z(t) = z_0 \cos \omega t$; s is the distance along the neutral axis of the beam; v and u are the horizontal and vertical displacements of the mass; and the unseen piezoelectric patches are placed along the beam.

cantilever beam with piezoelectric patches, loaded with a tip mass, was investigated. This chapter considers the electromechanical response of the pre-buckled inverted cantilevered beam, subjected to a combination of harmonic and broadband random excitations (see Figure 4.8).

The dependence "displacement–curvature" of the beam was nonlinear due to the large transverse displacement of the beam. The common idea of similar studies consists in adjusting the mass such that the system is near buckling and therefore has a low efficient resonance frequency. The beam is subjected to large strains demonstrating nonlinear behavior, that is, geometric nonlinearity. The exploitation of this nonlinearity is aimed to manufacture a low-frequency energy harvester, which is relatively insensitive to a particular excitation frequency and reacts on a sufficiently large amplitude.

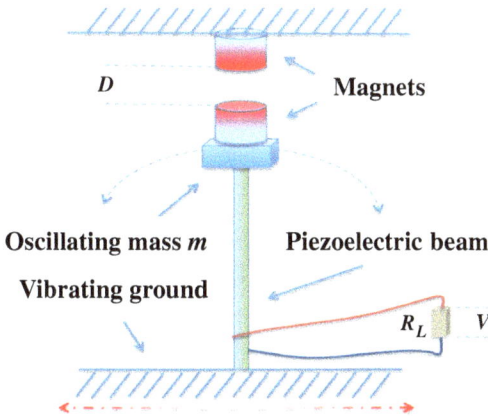

Fig. 4.9. Piezoelectric inverted pendulum, showing bistable dynamics.

A simple example, describing a process of transition from the linear (monostable) to a nonlinear (bistable) dynamics is shown in Figure 4.9. Let a small magnet (tip magnet) be added over the beam mass. During vibration, the pendulum oscillates alternatively bending the piezoelectric beam and generating a voltage V. The dynamics of the inverted beam tip could be controlled by an external magnet conveniently located at a certain distance D with opposed polarities to those of the tip magnet. In this case, a force dependent on the distance between magnets is introduced, which opposes the elastic restoring force of the bended beam. As a result, there are three types of behaviors of inverted pendulum dynamics in dependence on the distance D [113]:

(i) At $D \gg D_0$, the inverted pendulum behaves like a linear oscillator whose dynamics is resonant with a resonance frequency, determined by the system parameters (conventional piezoelectric harvester).

(ii) At $D \ll D_0$, the pendulum is forced to oscillate to the left or the right of the vertical. Taking into account small oscillations, it could be described again as a linear resonant oscillator with a resonance frequency higher that in the previous case.

(iii) At $D = D_0$, the pendulum demonstrates a more complex behavior with small oscillations around each of the two equilibrium positions (left and right of the vertical) and big transitions from one to the other.

Figure 4.10(a) corresponds to the case: $D \gg D_0$, where the inverted pendulum behaves like a linear oscillator.

In Figure 4.10(b), the same quantities of Figure 4.10(a) are present, with the only difference that now the two magnets are not far away from each other but at a certain distance $D = D_0$. In this case, the potential energy shows clearly two distinct equilibrium points separated by an energy barrier.

In Figure 4.10(c), the potential energy becomes even more bistable at further decreasing the distance between the two magnets $(D \ll D_0)$, and the barrier grows up to a point, in which the jumps between two minima become fewer and less probable. The displacement dynamics gets confined in one well, and it shows lower amplitude that reflects into a smaller voltage amplitude V.

As it has been shown in [113], the average output power (average value of V^2/R_L) obtained for this piezoelectric harvester was presented as a function of the distance D between magnets. It defined an optimal distance D_0, where the output power attained a maximum.

4.4.2. *Bistable harvesters*

As it has been pointed in Section 4.4.1, a nonlinearity is often introduced by creating two stable states in a harvester (bistability dynamics) [394]. Bistable oscillating harvesters have a unique double-well restoring force potential (see Figure 4.11). This provides three distinct dynamic operating regimes, depending on the input oscillation amplitude [149]. Bistable devices may exhibit low-energy intrawell vibrations (Figure 4.11(a)). In this case, the inertial mass oscillates around one of the stable equilibria with a small stroke per loading period. On the other hand, the bistable oscillator may be excited up to aperiodic or chaotic vibrations between wells (Figure 4.11(b)). At the

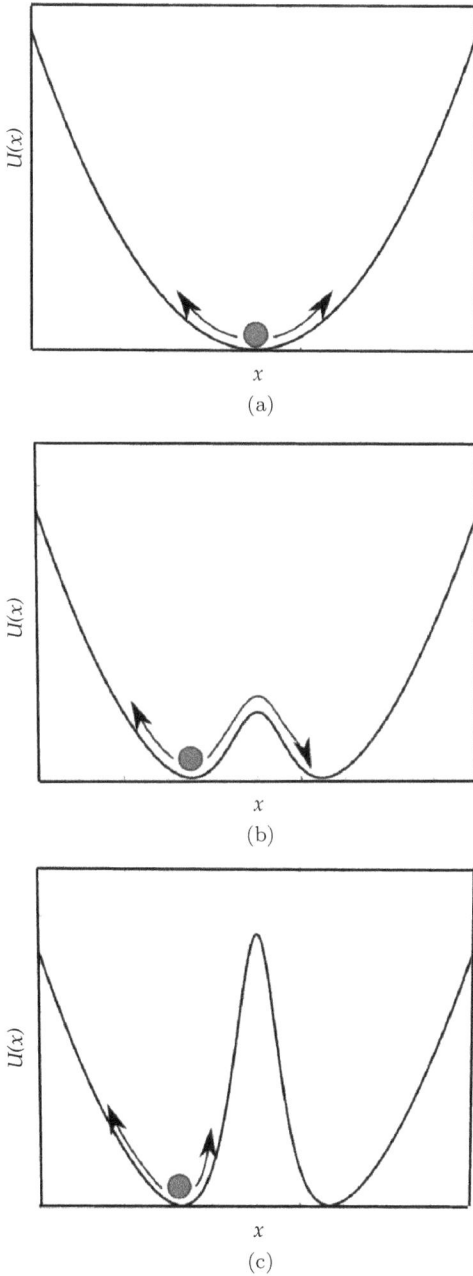

Fig. 4.10. Potential energy $U(x)$ in arbitrary units for: (a) $D \gg D_0$; (b) $D = D_0$; (c) $D \ll D_0$.

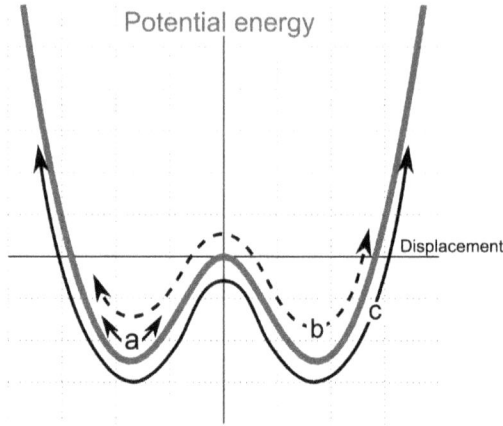

Fig. 4.11. Double-well force potential of bistable oscillator with instances of: (a) intrawell oscillations, (b) chaotic interwell vibrations and (c) interwell oscillations.

further growth of excitation amplitude, the device may demonstrate periodic interwell oscillations (Figure 4.11(c)).

As it has been shown in [409,503], the periodic interwell vibrations were the means in order to dramatically improve energy harvesting performance. Three general bistable harvester concepts are depicted in Figure 4.12. Figure 4.12(a) demonstrates a magnetic repulsion harvester with the strength of the nonlinearity governed by the magnet gap distance d_r. Figure 4.12(b) shows a magnetic attraction bistable harvester by using a ferromagnetic beam, directed toward one of two magnets separated a distance $2d_g$ from each other and d_a from the beam end. Figure 4.12(c) demonstrates an example of a buckled beam harvester with the bistability modified by a variable axial load p.

The different examples of bistable energy harvesters have been designed and experimentally investigated by various authors and characterized by corresponding bistability mechanisms.

4.4.2.1. *Magnetic repulsion bistability*

In [242], it has been demonstrated that the bistable harvester in Figure 4.12(a) consistently generated greater peak voltage than

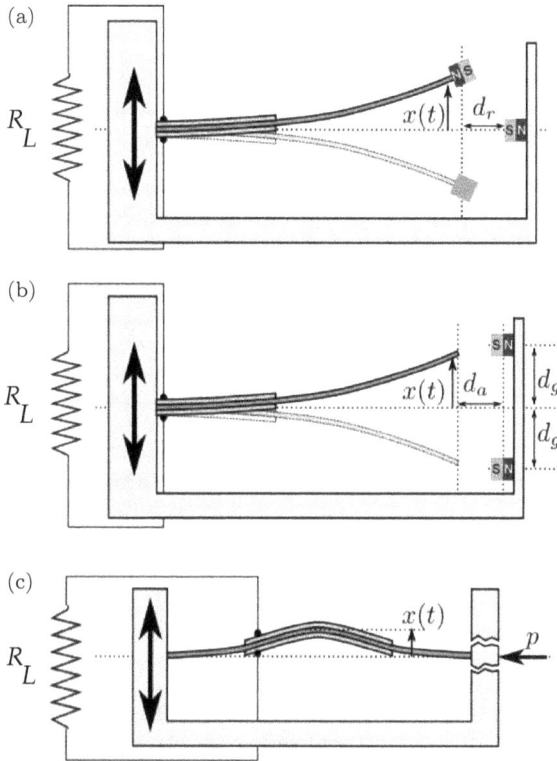

Fig. 4.12. (a) Bistable magnetic repulsion harvester; (b) magnetic attraction harvester; (c) buckled beam harvester: piezoelectric patches are shown as light gray layers along part of the beam lengths; R_L is the electric load resistor.

the equivalent linear device, when excited by pink noise. In this case, the bistable harvester provided 50% greater voltage than the linear device. An optimum magnetic repulsion gap was observed in [410], at which a considerable increase in broadband power could be harvested. The bistable harvester also provided 50% greater voltage compared to the linear device.

In order to create bistability, magnetic repulsion of a magnet, oscillating along the tube axis, was used [266]. A circular array of cantilevered piezoelectric beams with magnetic tip masses, activated by a vertical-axis windmill, having a shaft at the center of the beam array, was developed in [188].

4.4.2.2. *Magnetic attraction bistability*

The pulse snap-through was exploited as a conversion technique with increasing the frequency to excite linear harvesting devices [149]. In this configuration, a centrally suspended magnet was attracted by two end-suspended magnets along the tube axis (see Figure 4.12(b)).

The use of a bistable energy harvester to power a pacemaker [189] has shown its effectiveness in the range from 7 to 700 beats per minute when the harvester pacemaker used opposite permanent magnets to create bistability.

Thus, nonlinearity can be introduced by two attractive or repulsive magnets. Both monostable and bistable nonlinear configurations with different extents of nonlinearity were achieved by adjusting the gap between the magnets. When a certain amplitude is exceeded, the system jumps between the two states in a nonlinear, nonresonant, and chaotic manner. This lack of a well-defined resonance frequency leads to the fact that the device becomes effective in a wider frequency range. The optimal performance of the nonlinear energy harvester was achieved near the monostable-to-bistable transition point [408]. Both monostable and bistable configurations demonstrated significant advantages over the linear harvester near this point.

4.4.2.3. *Mechanical bistability*

Bistability, induced by an applied axial load, can be attained by using an inverted clamped piezoelectric beam and a tip mass selected to bend the system. This configuration was explored in [110] to demonstrate the advantages of the design for extremely low-frequency vibration media.

It is well-known that the post-buckled beam snaps from one stable state to the other at exciting by sufficient input loading. In energy harvesting, piezoelectric patches are coupled to the beam such that oscillations of the beam deform the piezoelectric layers (see Figure 4.12(c)). In [185], an array of linear cantilevered piezoelectric harvesters clamped to a post-buckled clamped beam was used. An optimum excitation frequency was measured and equal to approximately one-third of the natural frequency of the clamped cantilevers. Power harvesting near this excitation frequency was much less sensitive to frequency changes in comparison with the single cantilevers excited at their natural frequencies. It shows another advantage of bistable energy harvesting in providing a broad harvesting bandwidth.

As it has been shown in [159], the frequency bandwidth of a bistable piezoelectric harvester can be increased by 180% by using subharmonics, determined in bistable systems. In [227, 228], a modified bistable beam piezoelectric harvester was presented. The beam had an M-shaped bent steel spring with four piezoelectrics on the beam ends and a proof mass in the middle of the beam.

Optimal lay-up configurations and aspect ratios for energy harvesting were determined in [34] by using initial modeling and experimental studies, applied to illustrate the potential of the bistable harvester plate [17]. The test specimen in [17] was a nonlinear PEG, based on a bistable composite plate with bonded piezoelectric patches for broadband nonlinear energy harvesting. Bistability, in this case, was provided by the properties of the plate to retain its shape, but not by magnets. The obtained maximum output power was equal to 34 mW at pulsed excitation with a frequency of 9.8 Hz.

The bistable plates have significant advantages over bistable cantilever beams, such as a non-magnetic structure, which can mitigate the negative effects of the magnet used in electronics, thus is a more compact structure because of removing the magnets, and has more adjustability due to the two transverse dimensions of the plate.

The performed reviews of bistable energy harvesting [149, 323] have classified potential methods for creating bistability, such as using magnetic attraction/repulsion on cantilever structures and giving mechanical bistability to piezoelectric structures, for instance, by engineering asymmetric composite laminates supporting the piezoelectric structure [34]. In this study, a nonlinear bistable piezo-composite energy harvester was optimized using sequential quadratic programming. The design space was highly nonlinear and multimodal, but it was possible to consistently determine all local and global optima by using several random initial solutions. The dimensional parameters of the rectangular plate were optimized to maximize the output energy, characterized by maximum strain. Due to the nonlinear nature of the bistable structure, the strain is large and the output power can be as much as an order of magnitude greater than a linear harvester, with an additional benefit of harvesting the significant energy over a broad spectrum of excitation frequencies [46].

Finally, as it has been shown in [150], by using the continuous wavelet transformation, phase portraits, and multiscale entropy analysis, a comprehensive basis to characterize the dynamics and electromechanical behavior of bistable harvesters could be established.

4.4.3.   *Challenges in bistable energy harvesting*

Several key challenges in theory, experiment and modeling the bistable energy harvesters remain and are corresponded with some proposed solutions.

4.4.3.1.   *Maintaining high-energy orbits*

In [97], it has been demonstrated that a mechanical impact loading of the system could help the bistable harvester recover a high-energy orbit. Moreover, a sudden change in external circuit impedance could destabilize the intrawell vibration, returning the oscillator into a high-energy orbit, as it was observed in [272].

One of the main issues coupled with bistable piezomagnetic systems is the relatively high potential well, which restricts the application of such harvesters to high oscillation amplitude applications in order to provide enough energy to cross the potential well [499].

Understanding of the excitation characteristics, required to induce interwell dynamics, is a direction of rigorous mathematical investigations. Melnikov theory [393], period-doubling bifurcation [149] and evaluation of Lyapunov exponents [460] are promising methods to quantify the threshold between intrawell and interwell oscillations.

4.4.3.2. *Stochastic oscillating environment and coupled systems*

In [244], it has been shown that a specific noise intensity maximizes the harvested power obtained from bistable devices, having a static potential-energy profile. Therefore, if for an actual environment, a typical dominating stochastic oscillation is known, then the bistable harvester could be optimally designed. To illustrate the dynamics of phenomena occurring in a bistable system (in particular, in order to characterize the depth of the potential well), the phase portraits technique, based on Poincaré maps, was developed [38, 149, 214].

There is a great interest in the study of coupled systems, demonstrating chaotic behavior for the means of advantageous synchronization and array control [212]. The ambient noise is the source of loading, which can excite the resonant system stochastically into a bistable nonlinear response. Stochastic resonance could be induced only with moderate damping regardless of coupling strength. The possibility of coupling bistable harvesters has been considered in [245]. It has been indicated that identically excited bistable harvesters having various linear resonances could become unsynchronized.

Stochastic resonance, applied to energy harvesting, demonstrates three main features: (i) an energetic activation barrier proper for the double-well potential of a bistable system, (ii) a weak but coherent control input in the form of a periodic signal, and (iii) a source of ambient vibration that is inherent to the harvested system. The stochastic resonance could be considered as result of the synchronization of a stochastic time-scale, determined by the transition rate over the barrier, with a deterministic time-scale, determined by the time-scale of periodic modulation.

The presence of bistability makes the system capable of quickly switching between stable states. Even at relatively weak periodic excitation applied to the system, the double-well potential can periodically raise and lower the potential barrier. Therefore, noise-induced hopping between the potential wells can become synchronized with the periodic excitation, leading to stochastic resonance. Thus, it is necessary to propose and validate a design for advanced bistable harvesters, where the principles of stochastic resonance are used optimally and the efficiency of energy harvesting is maximized [495].

4.4.3.3. *Performance metrics*

There is no necessary consensus in the literature on the preferred means to evaluate energy harvesting performance [323], as the type of input excitation considered and the spectral bandwidth of relevance vary.

4.4.4. *Tristable and multistable harvesters*

In [498], a triple-well nonlinear piezoelectric energy harvester has been designed including a piezoelectric beam with a tip magnet and two external magnets with a unique arrangement, compared to conventional bistable systems. In their investigation, the authors presented a theoretical model and experimental studies of this PEG. Predictions of the theoretical model and experimental results in the frequency range from 10 to 35 Hz showed that a significant amount of energy can be harvested in the frequency range from 15.1 to 32.5 Hz under smaller oscillation amplitudes compared to conventional bistable systems.

The simplified diagram of the considered tristable piezoelectric energy harvester is shown in Figure 4.13. The magnet at the substrate layer end provided nonlinear forces due to interaction with two other magnets. The energy was stored by the energy harvesting circuits, operating on the base of the piezoelectric effect for the piezoelectric layers, moved by the substrate layer. A coupled electromechanical model with a nonlinear magnetic restoring force was used to define the dynamic parameters of a tristable nonlinear PEG. The characteristics of the corresponding linear energy harvesting system without the presence of magnetic force were obtained using genetic algorithms by searching for the minimum error between numerical modeling and experimental results. The main conclusion of this study was that, in comparison with the bistable PEG configuration, which has deeper potential wells, the tristable configuration is more easily excited and

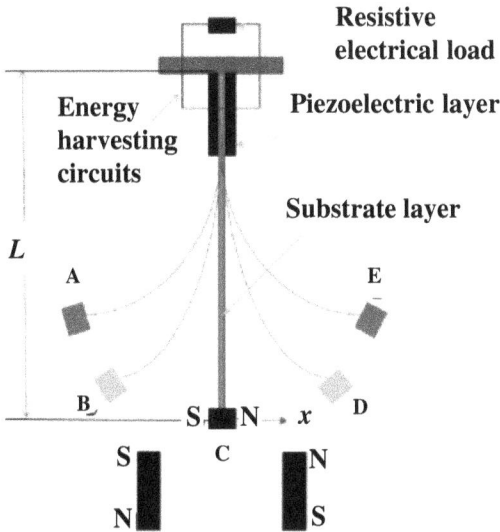

Fig. 4.13. Diagram of tristable piezoelectric cantilever energy harvester: X is the horizontal displacement of the substrate layer end at moment t; L is the length of substrate layer; B and D are unstable positions of the vibrating system; A, C and E are stable ones.

more easily overcomes potential wells. It allows one to attain higher output power in a broad frequency range.

Global dynamics of a piezoelectric energy harvester with tristable potential was studied in [504]. The dynamical model of a cantilever-beam energy harvester was considered with its static bifurcation. Based on the multiple intrawell attractors and their basins of attraction, the mechanism of multistability and its initial sensitivity were discussed. Moreover, the Melnikov method was applied to state the conditions of global bifurcations and the induced complex dynamics. The results showed that the variation of polynomial coefficients stated the number and shapes of potential wells. At the same time, increasing the excitation amplitude triggered multistability around one equilibrium, initial-sensitive jump, interwell attractor and chaos.

A multistable piezoelectric energy harvester with a nonlinear spring, subjected to wake-galloping, was developed in [234]. A piezo-electric cantilever-cylinder structure for energy harvesting caused by vortex-induced vibration was constructed in [292]. It was determined that the energy harvester in the monostable configuration displayed a hardening behavior with higher amplitudes. So a larger output voltage has been attained. At the same time, it had a wider synchronization region with period or non-period responses, but produced a lower output power in the bistable configuration. A magnetic levitation-based hybrid energy harvester was considered in [477]. In this case, the tristable system required less kinetic energy to excite a large displacement motion, compared with monostable systems.

An ultra-low-frequency energy harvester to harness structural oscillation energy was proposed in [444], displaying the preferences of multistability for energy harvesting. A double magnet system, including two piezoelectric beam energy harvesters with interacting tip magnets, was developed in [501]. The system demonstrated a multistable nonlinear behavior with a tunable frequency bandwidth via adjustment of the horizontal distance between the endmost magnets. The frequency bandwidth of the transducer depended on the linear parameters of every harvester without magnetic coupling, such as the horizontal distance between the harvesters and the size and characteristics of the tip magnets.

It is evident that multistability and chaos show great potential in vibration energy harvesting devices. These two initial-sensitive phenomena, namely chaos and jumps between intrawell motions and interwell ones, can easily create a large strain on the piezoelectric structure, collecting great electricity.

4.5. Piezoelectric Generators with Constructive L-Shaped Elements

PEGs with constructive elements of different shapes have been developed for improving output characteristics. We shall consider L-shaped elements in some detail. An L-shaped beam-mass construction was developed in [60] to study a possibility to use the nonlinear modal interaction found in the internal resonance for improving the frequency bandwidth of energy harvesters together with investigation of the influence of system parameters on the bandwidth.

It was concluded that the presence of the peak, connected with the two-mode component of the response, increases the bandwidth of the system, in comparison with a linear system. Moreover, calculation analysis showed that the frequency bandwidth was inversely proportional to the external load and piezoelectric electromechanical coupling factor.

It was demonstrated in [246], that the frequency bandwidth and output power are highly tunable by changing the length of the piezoelectric beams and the corner mass. A comparison between the studied nonlinear harvester and a cantilever beam harvester demonstrated that, at optimal transducer geometry, more output power could be harvested from the L-shaped beam, while the power density of a cantilever beam was much higher compared to the L-shaped harvester.

Another two-to-one internal resonance mechanism was introduced in [473] by using a tuned auxiliary resonator, coupled to a primary oscillating structure, equipped with a permanent magnet to provide nonlinear behavior. Application of the analytical model allowed one to study the nonlinear dynamics and energy harvesting performance of the designed device. A similar study of increasing the piezoelectric energy harvesting by using internal resonance in an L-shaped vibration energy harvester has been presented in [148].

A finite element and test approach to modeling vibrations of a new axial-type PEG with a proof mass and an active base has been considered in [64]. A pair of cylindrical PEs, disposed along the generator axis and clamped with an L-shaped bar, was used as an active base. Two bimorph plate-type PEs on an elastic PEG base used the potential energy of PEG bending vibrations (see Figures 4.14 and 4.15). Energy generation in cylindrical PEs took place, owing to the transfer of compressive load to the PEs at the PEG base at excitation of oscillations. The results of modal and harmonic analysis of oscillations were presented for the PEG output characteristics at a low-frequency loading. The maximum output power, attained for each cylindrical PE was equal to 2138.9 μW, and for plate-type PEs, respectively, 446.9 μW and 423.2 μW. More complete results for the axial-type PEG will be presented in Chapter 5.

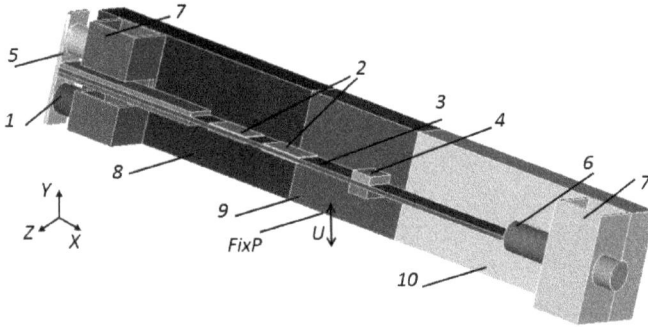

Fig. 4.14. Structure scheme of PEG with proof mass: (1) piezocylinder; (2) plate
PEs; (3) base; (4) proof mass; (5) L-shaped clamping bar; (6) rigid fastening of
the PEG right-end to the base 3; (7) fixing supports of PEG; (8–10) rigid base of
the generator; (9) PEG support part with applied displacement (*FixP*).

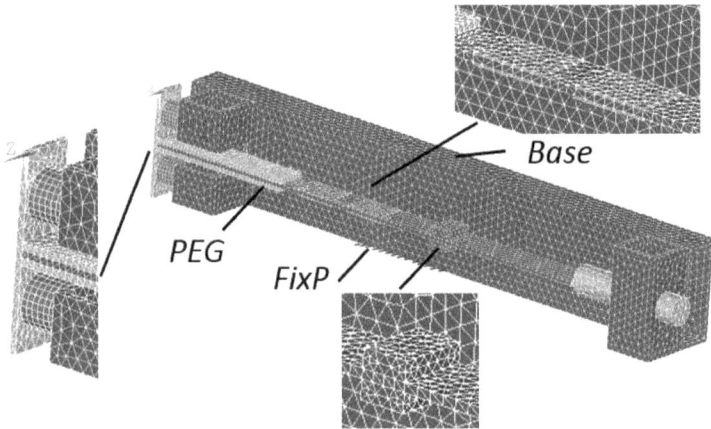

Fig. 4.15. FE-model of axial-type PEG: *FixP* are the nodes of rigid fixation of
displacement; *Base* — PEG fastening base.

The optimal design of an axial-type PEG has been presented
in [137]. The axial-type PEG had two piezoelectric cylinders d_{33}
(1), two piezoelectric bimorphs d_{31} (2), duralumin substrate (3),
proof mass (4), and L-shaped steel plates (5). One end of the
piezoelectric cylinders was fastened to the L-shaped steel plate and
the other end was fixed by a rigid pinch (6). Both the L-shaped
steel plates were fastened to the duralumin substrate. The proof

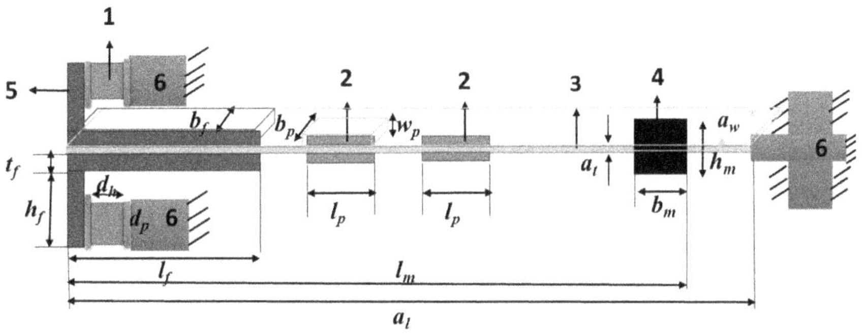

Fig. 4.16. Schematic diagram of the piezoelectric axial-type PEG with two L-shaped steel plates.

mass was fixed between two piezoelectric bimorph patches on the duralumin substrate. The two piezoelectric bimorphs were fixed on the duralumin substrate (see Figure 4.16).

The optimization problem was directed to obtaining the optimal output voltage for a given mechanical exciting load, as a result of which the PEGs had active and passive domains. This chapter was focused on the optimization process in the passive domain in order to increase the output voltage at given mechanical loading. The optimization was performed for specific operating conditions, different lengths of duralumin base plate, locations of proof mass and values of applied acceleration. The maximum output voltage and power were obtained equal to 11.64 V and 1355 μW, respectively, at a frequency of 633 Hz, when the length of the duralumin base plate was 150 mm and the acceleration was equal to 5 m/s^2.

4.6. Stack-Type Piezoelectric Generators

Two main kinds of cylindrical PEGs have been designed successively, namely, monolithic and stacked ones. The monolithic transducer presents itself a solid sample of piezoceramics with a certain shape and applied electrodes. A stack-type transducer consists of several monolithic cylindrical transducers, connected structurally into a single specimen, as well as coupled in parallel to a single electrical

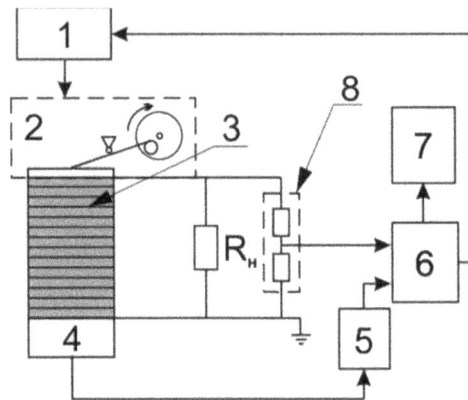

Fig. 4.17. Design scheme of the test set-up: (1) frequency transducer VFD004L21A, giving the frequency of mechanical load; (2) loading module; (3) test PEG; (4) strain gauge dynamometer; (5) voltage amplifier; (6) ADT/DAT; (7) PC with Power Graph software; (8) voltage divider.

circuit. The stack-type PEG configurations have been studied theoretically and experimentally in a number of studies (see, for example [237, 492]).

The authors of the book have participated in the development of an experimental approach to evaluating the output parameters of a stack-type transducer [6, 9]. The block-scheme of a low-frequency pulse mechanical load of PEG is shown in Figure 4.17. The laboratory set-up provides low-frequency mechanical loading of PEG with harmonic excitation oscillation in the modes of program and manual control of the amplitude and frequency components of the impact, acting along the PEG axis, with registration of input and output parameters of the mechanical load and response.

By modeling a whole harvesting system, a full equivalent circuit model was constructed in the Simscape/Simulink environment (see Figure 4.18). The main modules of a workflow diagram consist of the PZT stack, the full bridge rectifier, module of generating the random mechanical excitation, full electric load including battery and means for monitoring the process [357]. The experimental and model results of performed studies are present in some detail in recent review [319].

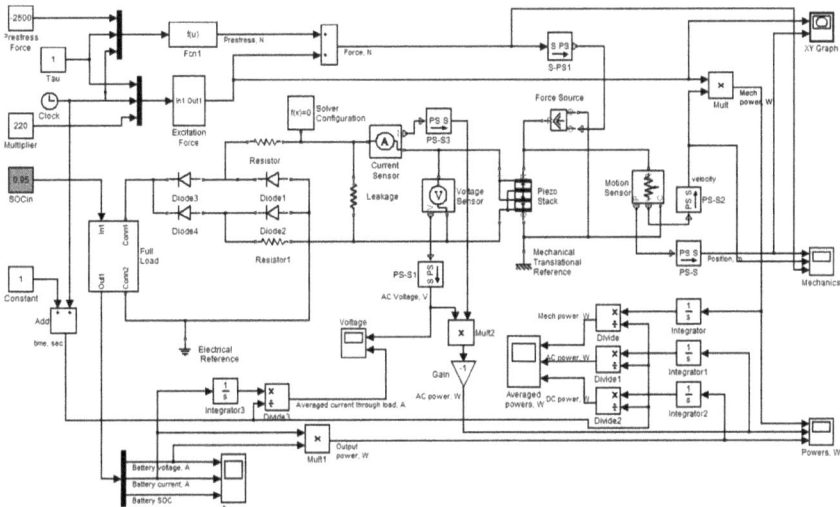

Fig. 4.18. Simscape–Simulink workflow diagram for PZT harvesting system.

4.7. Medical Applications

Continuing with the application of piezoelectric energy harvesting in cardiology, pointed by the framework of the discussion on the bistability phenomenon, we review in this section, some results, obtained in the last decade in drug delivery, blood sampling and ophthalmology.

4.7.1. *Piezoelectric micropumps with microneedles*

Piezoelectric micropumps have important and various applications in engineering and healthcare systems. The piezoelectric micropump has an accurate, precise volume flow, lower power consumption and controllable volume flow, which are most suitable for drug delivery and blood testing applications. Different designs of micropumps and microneedles that are suitable in these biomedical applications have been researched and developed by many scientists.

Microfluidics is the area in which various disciplines such as engineering, nanotechnology, biotechnology, biochemistry, physics, chemistry, etc. are merged to analyze the flow of low-volume fluids to reach automation, multiplexing and high-throughput transmission. The field of microfluidics started at the beginning of the 1980s and found applications for the enlargement of deoxyribonucleic acid (DNA) chips, inkjet printheads, micropropulsion, lab-on-chip technology, microthermal technologies, etc. The medical industry has shown an intense attention to microfluidics technology [78].

The micropumps play a significant role and are an important part of the microfluidic domain. The classification of micropumps is described in Figure 4.19 and it is focused on the diaphragm-based mechanical displacement micropumps. The piezoelectric diaphragm-based micropumps are used mainly in biomedical applications.

The micropumps are basically of two types, namely, valveless and check valve-based micropumps. For drug delivery applications, valveless micropumps are mostly used because of the congestion-free operation of diffuser/nozzle valves. Figure 4.20(a) shows a general mechanical displacement type valve micropump. It consists of an input valve, an outlet valve, a flexible diaphragm, a pumping chamber and an actuation chamber. Figure 4.20(b) demonstrates the valveless micropump of the nozzle/diffuser. The diaphragm is actuated by the corresponding mechanism; the up and down movement of

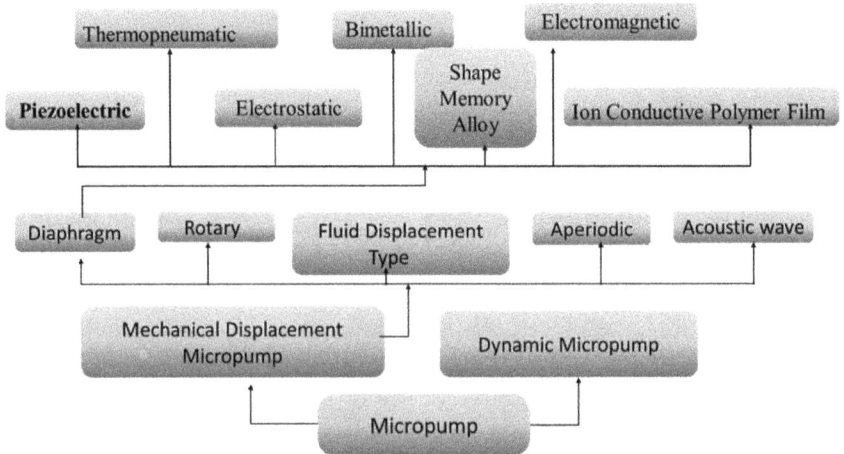

Fig. 4.19. Classification of micropumps.

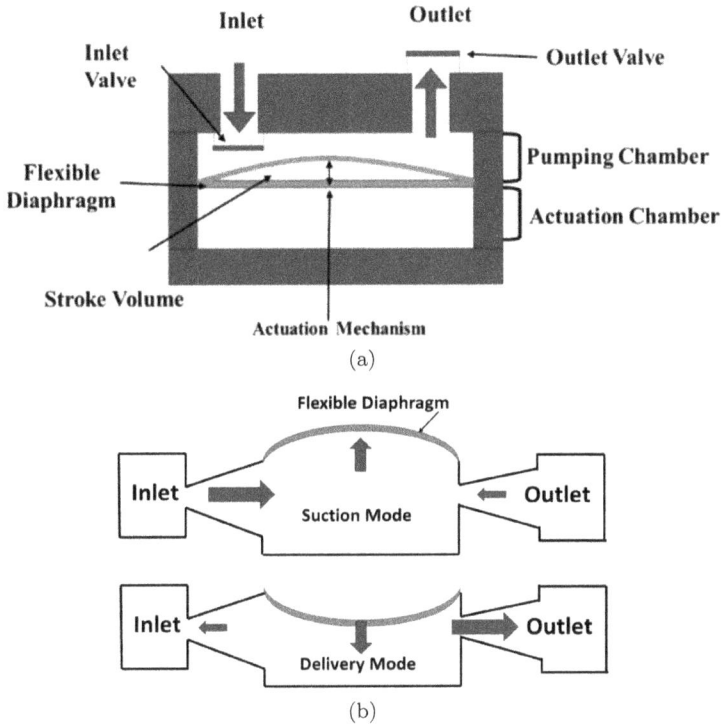

Fig. 4.20. Scheme of diaphragm-based micropump: (a) with valve and (b) without valve.

the diaphragm creates a pumping action. Various methods have been used by researchers for actuation of the diaphragm, namely, piezoelectric, thermopneumatic, shape memory alloy, electrostatic, electromagnetic, etc. [78]. Piezoelectric-actuated micropumps have the advantage of large actuation force, high stroke volume, and quick response. They are widely applied due to their simplicity in design. So, most biomedical applications are focused on piezoelectric actuation-based micropumps since the fluid flow through them is precise, accurate and controllable.

Many authors reported some basic guidelines for the selection of pumping principles, fluid chamber, backpressure, power consumption schemes and flow rate requirements in biomedical applications, such as blood transport and drug delivery. In [18], the microfluidic devices based on microelectromechanical systems have been reviewed for biomedical applications. The authors discussed the major features and issues related to microneedle and micropumps, working principles, actuation methods, fabrication techniques, failure analysis, construction, testing, safety issues, applications and future prospects.

The electronic mosquito has been designed for blood sampling, analysis and drug delivery applications. In this design, the PE is used for the actuation [140]. A series of disposable and individually activated e-Mosquito cells form the disposable patch. The array of 180° e-Mosquito cells provide periodic blood draws for up to 1 week, assuming that hourly blood monitoring is required. A compact human blood sampling device that can be used for the self-monitoring of blood glucose has been developed. This device has three main components: (i) a piezoelectric bimorph actuator, (ii) a microneedle control unit, and (iii) a painless and biocompatible microneedle. In [140], a wearable wristwatch for blood glucose measurement has been designed. The wristwatch has seven micropumps and each micropump would be used for a single blood sample. The piezoelectric micropump has been designed and simulated using ANSYS software to investigate the performance and behavior of the micropump. To improve the performance of the piezoelectric micropump, some important parameters, such as the actuator size, actuator shape, diffuser length, the diffuser angle and the neck width, should be considered [141, 144].

The disposable piezoelectric micropump for a closed-loop insulin therapy system has been designed, too. Through *in vivo* animal trials, the micropump has shown a possibility to control blood glucose. In [122], a micropump for drug delivery application with the flow rate of 20 μl/min and a maximum backpressure of 200 Pa has been considered. One application has great potential for micropumps based on polydimethylsiloxane (PDMS) in the treatment of glaucoma. The micropump was provided with variable active outflow and passive outflow aqueous humor [191].

Taking into account the environmental friendliness and low drug waste, an indirect drug delivery piezoelectric pump was designed.

Experimental results revealed that increasing the diameter of the needle increased the maximum output flow rate of the micropump. The volume flow rate of 0.96 ml/min was obtained at 160 Vpp and 24 Hz with a microneedle diameter of 0.3 mm [61].

The microneedle plays the role of an interface between the micropump and the human body. Microneedles are classified into three categories, namely, solid, hollow and dissolvable microneedles. Several tip geometries, such as a volcano, snake fang, microhypodermis, have been reported in the literature for microneedles [19]. The body or shaft of the microneedle, however, was formed into simple shapes such as conical, cylindrical, bevel or pyramid. Other geometries such as extremely sharp and side-open microneedles have also been reported [40]. In [232, 233], an ultra-high-aspect ratio (with 2 mm length and 60 μm inner diameter) microneedle and PDMS elastic self-recovery actuator have been introduced. The authors also studied the liquid extraction ability of the micropump system and optimized the microneedle for minimally invasive blood extraction. They observed the effect of bevel angle on the blood extraction time and reported that the bevel angle does not affect the extraction time, when the diameter of the microneedle is greater than 100 μm.

In [133], the effects of microneedle insertion depth, needle retraction, and infusion flow rate on associated pain have been studied. The shorter needles and slower flow rate resulted in low pain. A series of microneedles for drug delivery without pain has been used and the size of the microneedles was determined based on the anatomy of skin. The effect of the microneedle diameter in the flow of the micropump has been analyzed in [143]. In [331], an analytical model has been developed for improving the design of microneedles by minimizing insertion pain and increasing mechanical strength. The flow analysis inside the micropump has been carried out at various shapes of the microneedle. The simulation has been performed with Newtonian and non-Newtonian fluids [139]. Different researchers reported that pain is felt at the time of insertion of the microneedle into the body, therefore, the needle size should be as small as possible.

4.7.2. *Finite element modeling of lensotome tip with piezoelectric drive*

Another medical application of piezoelectricity is the operation instrument with piezoelectric drive, which is used in the lens replacement surgery (see Figure 4.21). In [378], the mathematical and computer modeling of the tip of a lensotome was carried out. The lensotome was presented by a hollow cylinder that vibrated axially and its oscillations were excited by a piezoelectric drive. A device for a lensotome tip with a piezoelectric drive consisted of elastic piezoelectric pads and contacts with the acoustic medium. The linear equations of the theory of elasticity and electroelasticity, taking into account the dissipation of energy, adopted in the ACELAN and ANSYS software, as well as the equations of motion of liquid and gaseous media in the acoustic approximation, were used as a continual model. A sufficiently complete and consistent presentation of the FEM for piezoelectrics and its implementation in the ACELAN software exists in [372]. The FEM of the lensotome tip in ACELAN software is shown in Figure 4.22, where a fragment of the finite element mesh is present.

The mathematical and FE models [378] were built for two cases of boundary conditions at the lower end of the tool: (i) free end and (ii) rigid fixation. In both cases, natural frequencies and coefficient of electromechanical coupling were calculated and natural vibration modes found. An analysis of these results showed that in the

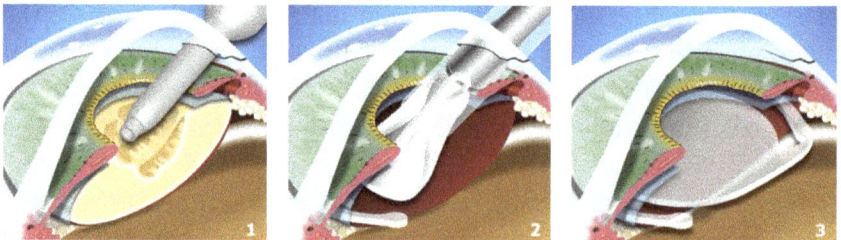

Fig. 4.21. Stages of the lens replacement surgery: (1) removal of clouded lens (phacoemulsification); (2) installing an artificial lens instead of removed lens; (3) implantation of intraocular lens (IOL).

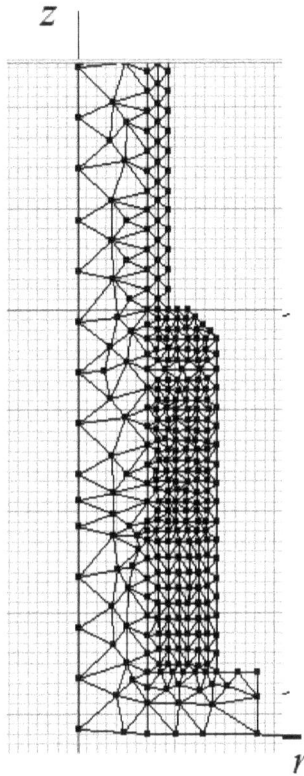

Fig. 4.22. Fragment of the finite element partition of the tip.

first case, the first mode was preferable, taking into account the restrictions on the frequency value. In the second case, the first and second modes were preferable. At the same time, the defined higher effective frequencies and vibration modes did not satisfy the limitation on the value of the operating frequency.

4.8. Conclusions

Piezoelectric energy harvesting has become one of the extremely in-demand research areas in the 21st century. Obviously, it is very difficult to summarize in a brief chapter all the achievements in this important area of scientific and technical research. However, this review chapter comprehensively presents existing designs and promising problems in the field of piezoelectric energy harvesting over the last decade. Multifunctional energy harvesting technologies have been developed by using cantilever-type, axial-type and stack-type PEGs and piezoelectric actuators with complex structures and geometries. Piezoelectric energy harvesting has found significant and numerous applications in medicine. Future developments of complex devices and networks including jointly operating piezoactuator and piezogenerator structures will be able to increase efficiency of energy harvesting and improve necessary output characteristics.

Continuous research and development of advanced piezoelectric materials and composites with improved electromechanical, thermal and biocompatible properties, including flexoelectrics, functional gradient materials, porous ceramics, lead-free and high-temperature piezoelectrics are at the center of investigations on energy harvesting, These promising materials and the transducers based on them have been applied widely in the development of devices that use energy harvesting, including infrastructure objects and vehicles, wind streams and liquid flows, the movement of human bodies and animals. There is a tremendous interest in the creation of different embedded energy harvesting devices and rotary harvesters for structural health monitoring in aviation applications.

Great attention should be devoted to using different nonlinearities (including material properties) for developing broadband energy harvesters based on the phenomena of monostability, bistability, tristability and multistability. Corresponding mathematical approaches and methods could find their important applications for solutions to these problems.

Finally, evidently significant progress over the last decade in the development of computer software has allowed one to raise the state-of-the-art of computer modeling of new complex energy harvesting devices, structures and systems to a new level, and will continue to do so in future.

Chapter 5

Experimental and Finite Element Modeling of Cantilever-Type and Axial-Type Piezoelectric Generators with Active Elements

5.1. Introduction

This chapter is devoted to a more detailed consideration of some piezoelectric power generation devices presented in Chapter 4.

Today, heat, light, radiation, wind and water are used as sources of environment energy, converted through local devices into electrical energy. The use of such resources as the vibration of environmental elements as a source of additional energy makes it possible to expand the range of elements on the basis of which additional electrical energy can be generated.

Devices for generating electrical energy, so-called power converters using piezoelectric elements, are called piezoelectric generators (PEGs). The development of various power generation devices requires their further modernization and improvement. This is possible only by constructing new, excellent devices and analyzing their operations under various dynamic loading conditions.

Basic information about energy generation, as well as problems arising in the development of energy storage devices using piezoelectric elements, were discussed in Refs. [12, 66, 149, 182, 202, 247, 286, 319, 322, 323, 341, 354, 448, 465]. Various periods of development of renewable energy devices are presented in Refs. [182, 202, 247].

The primary analysis of review papers shows that models are considered that use piezoelectric elements operating in compression and buckling with tension. Piezoelectric elements can be embedded into various mechanisms using both pressure and rotational loads. Moreover, the use of power generation devices can be based on the use of magnetic and magnetoelectric elements. References [98, 328, 358] present theoretical and experimental works of some authors. Methods of studying the devices with piezoelectric elements are given.

To solve the problems of harvesting energy by generators of various types, devices are being developed that have elements of autonomous power supply in their design. The most relevant in terms of their development and introduction into production are renewable electrochemical batteries. These electric batteries have the properties of cyclic recovery during a certain resource period and are limited in use when a certain finite operating time is reached [215, 478].

Some approaches to scavenging for environment energy that can be used as electric energy sources include the following: (i) the use of solar energy obtained by solar panels [132, 241]; (ii) the energy of the air streams (wind loads) [147, 337] harnessed with the help of piezoelectric energy generation converters [299, 466]; (iii) mechanical energy of the movement of water fluxes in narrowed volumes, sea tides, water flows of rivers and dams [47]; (iv) heat energy [452] and (v) mechanical vibrations of structural elements and soils [457, 463].

The devices, using elements of energy generating structures in the form of piezoelectric elements, as components of systems for converting mechanical energy into electrical energy require further development owing to low output power and generated voltage [51, 65, 100, 238, 381, 383, 384]. In Ref. [386], an energy harvester has been optimized and finite element (FE) analysis was carried out with the help of ANSYS and ACELAN software. The optimal design was based on matching the resonant frequency of the device with the exciting environmental frequency. The stack-type piezoelectric energy harvester has been investigated in Ref. [89]. An experimental set-up was designed to observe the response of multilayer piezoelectric stacks in the energy harvester. The single-degree freedom fractal structure system has been also proposed for the energy harvesting in Ref. [213]. The authors carried out a design optimization and experimentally evaluated the performance of the system.

The amplitude limiting rotational piezoelectric energy harvester has been designed to avoid resonance conditions [453]. Radial magnetic force was used for energy harvesting. The harvesting of piezoelectric energy has been reviewed and analyzed in Ref. [239] based on the characteristics of compliant mechanisms. The authors have classified the harvester configurations into mono-stable, multi-stable, multi-degrees-of-freedom, frequency up-conversion and stress optimization. They also introduced the normalized power density to compare the energy generation capability of energy harvesters.

The energy harvester to extract energy from smart road has been designed in Ref. [476]. The authors investigated the number of stack layers, influences of connection mode, number of units and ratio of height to cross-sectional area. Several studies considered various designs of piezoelectric generators in dependence on the area of application. By this a direct piezoelectric effect was used when the sensitive element oscillated longitudinally (d_{33}), under bending (d_{31}) and sharing [142,194–196,287,300]. In Ref. [196], a three-dimensional finite element analysis is presented for a cantilever plate structure, excited by piezoelectric drive sections. The paper considers modeling the actuators of optimal configurations for selective excitation of the modes of a cantilever plate structure. Such elements can be used for technical analysis of vibrations of various structures using micro electro-mechanical systems (MEMS) technologies [196,300].

5.2. Experimental Modeling

5.2.1. *Description of model*

Figure 5.1 shows a model sample of cantilever-type PEG, consisting of an elastic material on which the piezoelectric elements (5) are glued on both sides (bimorph), one end of the console (3) is fixed in the base (1), and the proof mass (4) is fixed on the free end, the base additionally has four piezoelectric elements (8), two at the top and two at the bottom relative to the cantilever beam (4) having the polarization vector directions The planes of the electrodes

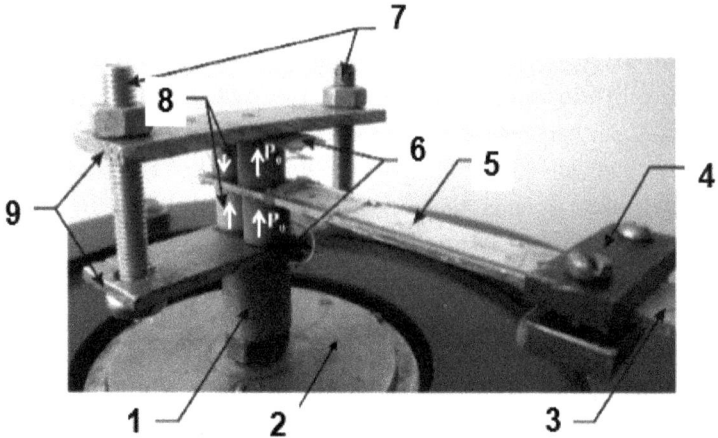

Fig. 5.1. Model sample of cantilever-type PEG in experimental set up: 1 — support stand; 2 — base of the vibrating table; 3 — plate of cantilever for fixing bimorph; 4 — proof mass; 5 — piezoelectric plate (bimorph); 6 — elastic insulating gaskets; 7 — tightening bolts; 8 — piezoelectric cylinders at the PEG base; 9 — pressure plates.

are pressed with the help of the elements (9) of the base (1) to the conductive layers of thin elastic gaskets (6), metallized on one side. The top piezoelectric cylinders have mixed polarization, the lower piezoelectric cylinders have polarization directed in the same direction and collinear along y-axis.

The proposed device works as follows. Under the excitation of piezogenerator base (1) by external mechanical shock-type forces and oscillations of vibration table (2), the flexural vibrations occur in the cantilever beam (3). They act on the piezoelectric elements (8), in which the variable compression strains arise due to the force reactions of the supports with the frequency of external shock and vibration forces. Owing to the direct piezoelectric effect, alternating current (AC) voltage generates on the electrodes of additional piezoelectric elements (6), creating additional electric energy. The combined use of such elements can improve the output power and conversion efficiency of the converter. This AC voltage and additional electrical energy can be converted by means of bridge rectifiers into direct current (DC) electrical energy, which is accumulated in the batteries by means of energy storage systems. The excitation of oscillations in the cantilever beam can be performed by mechanical action both on the base (1), on which the cantilever beam (3) is clamped, and on the free end of the cantilever beam (3). The maximum output power is achieved when the frequency of the external mechanical action coincides with, or is close to, the natural frequency of the layered cantilever, that is has a resonance value.

Piezoelectric elements can be connected in parallel, in series or have a separate connection. The choice between the types of coupled elements depends on the device that is necessary to be powered: if greater output voltage is required, a serial connection is selected, and if a greater output current is needed, then the parallel connection is selected.

Figure 5.2 presents a vibration set-up for the study of oscillatory processes of PEG. As the devices recording displacements, laser sensors of displacements optoNSDT (3) and RF603 (6) are used. The sensor (3) records the displacements of the PEG base. The RF603 (6) sensor was used to record the movements of the proof mass fixed

Fig. 5.2. Vibration set-up of PEG: 1 — vibrating table; 2 — PEG; 3 — laser sensor optoNSDT of displacements; 4 — mounting of the laser sensor 3; 5 — electric tract of PEG; 6 — triangulation laser meter RF603 of displacements; 7 — support column of the sensor 6; 8 — cylindrical-type PE, located at the PEG base; 9 — PE plates.

at the free end of the PEG cantilever. The following characteristics were recorded: voltage supplied to the base plug of the vibration table (1), output voltage from piezoelectric elements of PEG (8) and (9), passing the electrical path (5) to the analog to digital convertor (ADC), and base vibration signals from the transducers optoNSDT (3) and RF603 (6).

Figure 5.3 schematically shows a measuring system for the research of the output parameters of PEG under vibration excitation. Oscillations and output characteristics of PEG were investigated in stationary vibration excitation of vibration table. The obtained first resonant frequency of the generator was ∼22 Hz. Then, the signal in the form of a sine wave with a frequency of the first resonance was excited on the generator AFG3022 and a vibrating table (2) with a PEG, attached to it, was actuated through the amplifier (9). In this case, all piezoelectric elements were loaded with active resistance. Through matching devices (4), (6), (8), the ADC (7) received signals

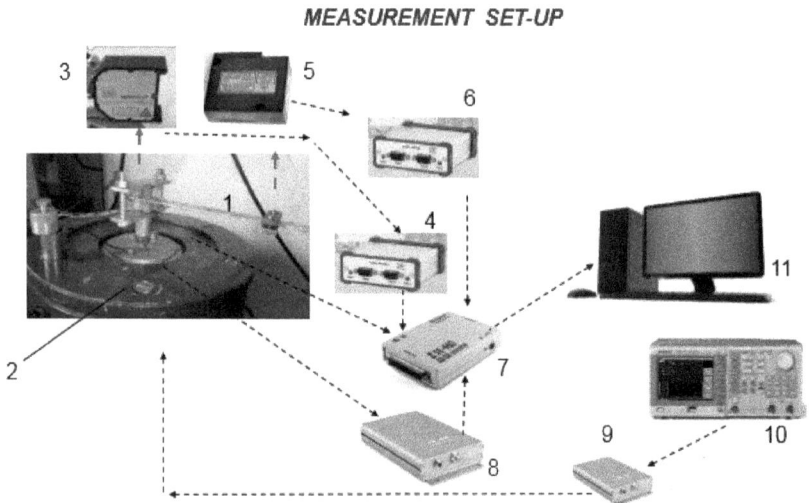

MEASUREMENT SET-UP

Fig. 5.3. Measurement set-up: 1 — PEG; 2 — vibration exciter; 3 — optical sensor optoNSDT of linear displacements; 4 — controller of the optical sensor 3; 5 — optical sensor RF603 of linear displacements; 6 — controller of the optical sensor 5; 7 — external ADC/DAC module L-Card14-440; 8 — matching device of the acceleration sensor; 9 — power amplifier; 10 — sets of the signal generator AFG3022; 11 — computer.

generated by the PEG and laser motion sensors and was recorded using software of a computer (11).

5.2.2. Results and discussion

Analysis of the voltage dependence on the active load shows that the voltage increases smoothly to the value of $U_{\text{Bim}} = 5.13\,\text{V}$ at a load equal to $2\,\text{M}\Omega$ for PE plates, located on the substrate (see Figure 5.4). For the upper piezocylinders with mixed polarization, the value of the generated voltage is greater than for the lower piezocylinders with the polarization directed in the same direction. These values are equal to $U_{PC\text{up}} = 3.66\,\text{V}$ and $U_{PC\text{dn}} = 2.61\,\text{V}$, respectively, under an electric load resistance of $2\,\text{M}\Omega$ (see Figure 5.4).

Analysis of the values of output power (see Figure 5.5) shows that for this design and fixing the PEG elements, when the proof mass is located in the extreme right position $(L_m = 150\,\text{mm})$, the peak power value of $P_{\text{Bim}} = 41.8\,\mu\text{W}$ and the output voltage of $U_{\text{Bim}} = 3.36\,\text{V}$ are achieved at an active load of $R = 270\,\text{k}\Omega$. The amplitude of the oscillations of the PEG base is 0.022 mm; the oscillation amplitude of the PEG end with proof mass reaches 0.67 mm at the first resonance 22 Hz. For the upper cylinders, a peak power of $P_{PC\text{up}} = 7.42\,\mu\text{W}$ and an output voltage of $U_{PC\text{up}} = 4.83\,\text{V}$ are achieved at a load

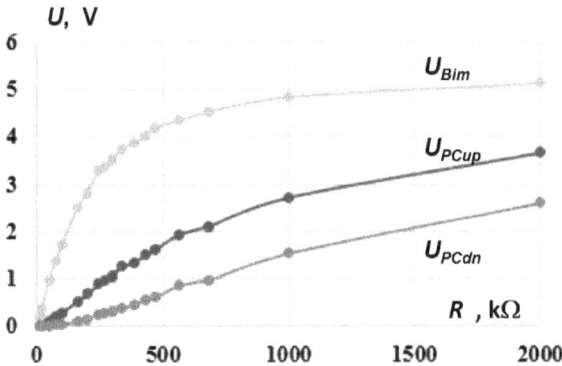

Fig. 5.4. Dependence of voltage on load impedance for the first mode of oscillation: U_{Bim} is the voltage on the bimorph, located on the PEG cantilever, $U_{PC\text{up}}$, $U_{PC\text{dn}}$ is the voltage on the electrodes for the upper and lower piezocylinders, respectively, located at the PEG base.

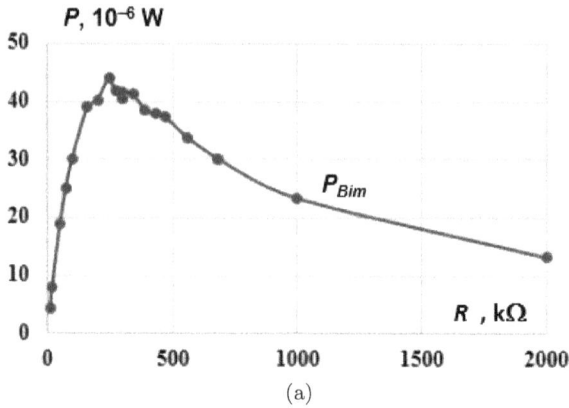

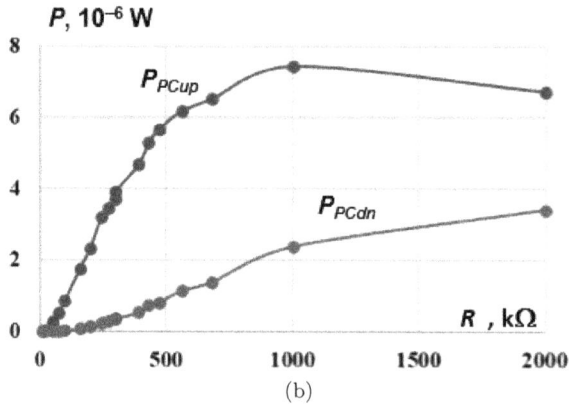

Fig. 5.5. Dependence of output power on load impedance for first mode of oscillation: P_{Bim} is the output power on bimorph, located on the PEG cantilever; P_{PCup}, P_{PCdn} are the powers on the electrodes of upper and lower piezocylinders, respectively, located at the PEG base.

of $R_{PCup} = 1\,M\Omega$. For the lower piezocylinders, a peak power of $P_{PCdn} = 3.41\,\mu W$ and an output voltage of $U_{PCdn} = 2.61\,V$ are attained at a load of $R_{PCdn} = 2\,M\Omega$.

5.3. Measuring Set-up with Rotating Drive

5.3.1. *Formulation of the problem*

Then laboratory set-up was developed for full-scale modeling of the cantilever-type PEG with proof mass and active base. Excitation of the PEG vibrations was carried out by a rotating drive having magnetic devices on its rim. Energy is generated due to the transfer of attraction forces to the magnet during rotation of the rotor, to the proof mass and owing to the bending of the PEG base.

The use of such a PEG requires the delivery of vibration energy to two of its points: (1) to the leg of the PEG base and (2) the point located on the generator plate. In this case, the greatest bending moment will arise at the edge of the base plate. The use of magnets in power generation devices can also serve as a basic element for loading a generator. So approaching a magnet to the metallic proof mass can create its impulse loading due to attraction or repulsion forces. Mounting magnets on wheel drives will lead to an oscillatory displacement of the cantilever base plate due to the appearance and disappearance of magnetic forces during the rotation of the wheel.

5.3.2. *Test set-up*

Figure 5.6 shows the test set-up with the PEG loading drive. It consists of a drive wheel (7) with a fixed C25-type magnet (4) with a diameter of 25 mm. The number of magnets on the set-up may not be

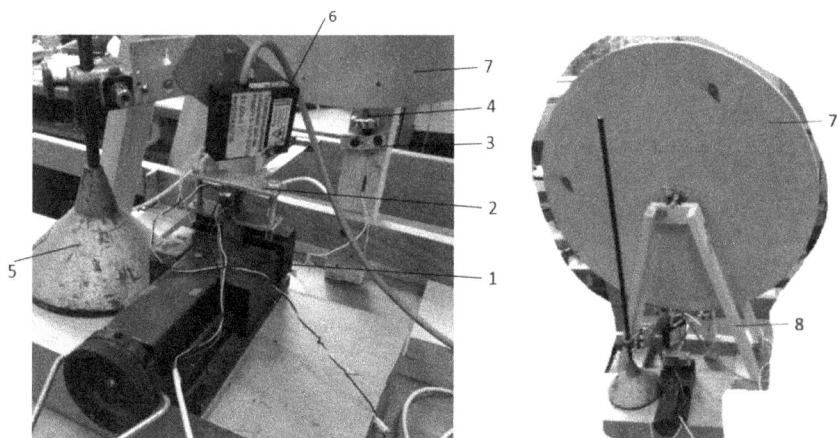

Fig. 5.6. Vibration set-up for PEG excitation: 1 — clamp; 2 — PEG; 3 — inductive sensor; 4 — magnet; 5 — fastening elements of laser sensor RF603; 6 — laser triangulation displacement meter RF603; 7 — rotating drive; 8 — elements of drive base.

limited and depend on the research goals. In this variant, only one magnet was attached to the wheel. The drive is mounted on the rod construction (8) of the base on a bearing support, and the PEG is fixed in the clamp (1). To measure the displacements of individual points of the base, a laser displacement meter RF603 was used. It detects the oscillations of the PEG cantilever in the vicinity of the extreme point of the bimorph attachment. The drive wheel with a diameter of 67 cm is made of plywood with 12 mm thickness. Inductance sensors, located on both ends of the PEG, allow recording the time of the passage of the magnet. The angle between two sensors relative to the center of the drive wheel is equal to 0.3795 rad. The distance between the magnet and the proof mass (located in the region of the free end of the generator cantilever at a distance of 150 mm) was chosen equal to 7 mm.

5.3.3. *Description of the work set-up*

The set-up operated as follows (see Figure 5.7). During rotation of the drive wheel (2), oscillations and output characteristics of PEG (1) were studied. At the wheel (2) rotation, the fixed magnet (5) excited an electric potential, when it passed near two inductive sensors (3, 4), located before and after the proof mass (1). Thus, the excitation time of the first and second inductive sensors was fixed and the average rotation speed was calculated. When the magnet passed over the proof mass, the metal mass was attracted to the magnet. After passing the magnet, and thus removing the attraction, PEG continued to oscillate with a certain damping. The output excitation voltage of the induction sensors and the values of PEG output voltage, taken from the bimorph and piezoelements in the PEG base, were recorded. In this case, all piezoelectric elements, according to Figure 5.7, were loaded with active resistance $R = 1\,\text{M}\Omega$. With the help of matching devices (1.1), (1.2) and (7), the ADC (8) received signals, generated by PEG (1) from induction sensors (3), (4) and a laser displacement sensor (6), with their subsequent recording using software on the computer (9).

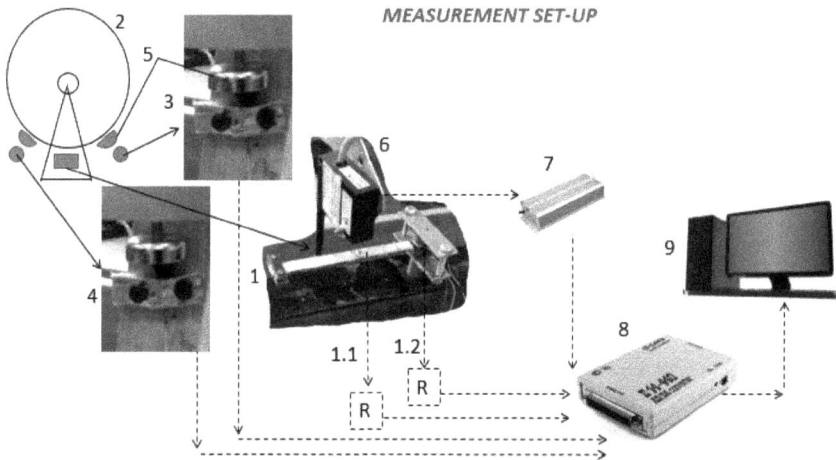

Fig. 5.7. Schematic representation of the measuring set-up: 1 — PEG; 1.1, 1.2 — transmission path from the bimorph and cylindrical PE; 2 — drive wheel; 3, 4 — induction sensors before and after PEG for fixing the position of magnets; 5 — magnet; 6 – laser displacement meter RF603; 7 — controller; 8 — ADC module L-Card14-440; 9 — PC.

5.3.4. *Results and discussion*

The studies were carried out using the Powergraf software, which makes it possible to register signals from sensors and PEGs and record them on a computer. Examples of test signals, read from displacement sensors (6) and piezoelectric elements (1.1, 1.2) in Figure 5.7, are shown in Figure 5.8. Plot 1 of the movement of the PEG base was built by using a laser displacement sensor RF603. The displacements of the PEG base were recorded at a point in the vicinity of the attachment of the piezoelectric plates. Plots 2 and 3 show the voltage on the inductive sensors, located before and after the PEG. An electrical voltage was generated when the magnet passed in the vicinity of the sensors. The difference between the voltage peaks allowed one to calculate the speed of movement of the drive wheel. Plot 4 shows the voltage, generated on the PEG base bimorph. Plots 5 and 6 show the values of voltage, excited on cylindrical piezoelectric elements. In this case, the results for the

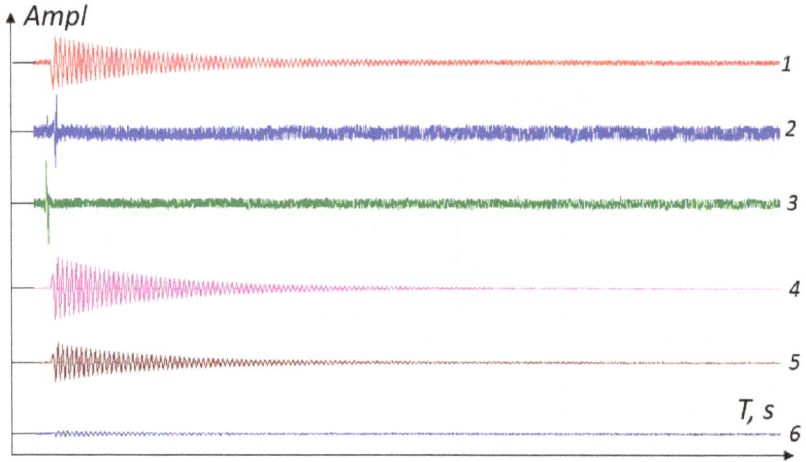

Fig. 5.8. Testing the PEG operation at excitation by magnetic drive: 1 — displacement amplitudes of the PEG base; 2, 3 — voltages on the electrodes of induction sensors; 4 — voltages on bimorph PEs at PEG base; 5, 6 — voltages on cylindrical Pes.

voltage on the electrodes with co-directed polarization piezoelectric elements (PEs) are shown in plot 5, and plot 6 corresponds to the oppositely directed polarization PEs. The damped oscillations were formed at the first natural frequency of the PEG.

Further, experimental testing of the PEG was carried out at its loading with help of a magnetic drive at low speeds. The rotation of the drive wheel was performed at a speed from 2.98 to 4.92 rad/s, which corresponded to the speed of rotation of the wheel rim from 0.98 to 1.62 m/s. The values of peak voltages (U_1, U_2, U_3) during harmonic excitation of PEG by a magnetic drive depending on the speed of rotation of the drive wheel are present in Table 5.1. For bimorph and cylindrical PEs, the output voltage increased linearly depending on the drive rotation speed and was $U_1 = 3.09\,\text{V}$ on bimorph electrodes, and $U_2 = 0.47\,\text{V}$ and $U_3 = 0.04\,\text{V}$ on cylindrical PEs electrodes at active load $1\,\text{M}\Omega$ and rotation speed of $4.92\,\text{rad/s}$. Corresponding values of output powers, depending on the speed of rotation of the drive wheel, are also presented in Table 5.1. The total power was measured for $5\,\text{s}$, beginning from the excitation moment of PEG oscillations. The maximum power during this time for the

Table 5.1. Peak values of output voltage and power on bimorph and two cylindrical PEs at harmonic excitation of PEG by magnetic drive in dependence on the rotation speed of drive wheel.

Drive wheel rotation speed, rad/s	Output voltage U, V			Output power P, μW		
	U_1	U_2	U_3	P_1	P_2	P_3
2.9832	0.9388	0.1497	0.01497	0.3574	0.0087	0.00087
3.3345	1.685	0.259	0.0259	1.3452	0.0331	0.00331
3.908	2.3537	0.3628	0.03628	2.6721	0.0649	0.00649
4.9218	3.095	0.4748	0.04748	4.5548	0.1096	0.01096

corresponding PEs was $P_1 = 4.55 \times 10^{-6}$ W, $P_2 = 0.11 \times 10^{-6}$ W and $P_3 = 0.01 \times 10^{-6}$ W.

5.4. Modeling the Composite Elastic, Electroelastic and Acoustic Media by Finite Element Method

The energy harvesting PEG is a composite elastic and electroelastic body. It is assumed that the device performs elastic small oscillations in a moving coordinate system. The rectilinear vertical motion of this system in the area of fixation is given by the law for steady oscillations, as follows:

$$u(t) = \bar{u}e^{i\omega t}. \tag{5.1}$$

In this case, an adequate mathematical model of the operation of the device is the initial-boundary value problem of the linear theory of electroelasticity [208].

In the common formulation, the constitutive equations for a piezoelectric medium are written as

$$\rho\ddot{u}_i + \alpha\rho\dot{u}_i - \sigma_{ijj} = f_i,$$
$$D_{i,i} = 0,$$
$$\sigma_{ij} = c_{ijkl}(\varepsilon_{kl} + \beta\dot{\varepsilon}_{kl}) - e_{ijk}E_k,$$
$$D_i + \varsigma_d\dot{D}_i - e_{ikl}(\varepsilon_{kl} + \varsigma_d\dot{\varepsilon}_{kl}) - \epsilon_{ik}E_k, \tag{5.2}$$
$$\varepsilon_{kl} = (u_{kl} + u_{lk})/2,$$
$$E_k = -\varphi_k,$$

where ρ is the material density, u_i are the components of the displacement vector, σ_{ij} are the components of the stress tensor, f_i are the components of the vector of the density of mass forces, D_i are the components of the electric induction vector, c_{ijkl} are the components of the fourth rank tensor of the elastic moduli; e_{ikl} are the components of the third rank tensor of piezoelectric coefficients; ε_{kl} are the components of strain tensor, E_k are the components of the electric field vector, φ_k are the components of the electric potential, ϵ_{ik} are the components of the dielectric constants tensor, α, β, ζ_d are nonnegative damping coefficients (the value of ζ_d is used in ANSYS software).

For elastic medium, we have

$$\rho \ddot{u}_i + \alpha \rho \dot{u}_i - \sigma_{ijj} = f_i,$$
$$\sigma_{ij} = c_{ijkl}(\varepsilon_{kl} + \beta \dot{\varepsilon}_{kl}), \qquad (5.3)$$
$$\varepsilon_{kl} = (u_{kl} + u_{lk})/2.$$

Since harmonic analysis is used in the calculations, the following actions are performed for the corresponding components of the equations:

$$\dot{u} = i\omega u,$$
$$\ddot{u} = i\omega u^2. \qquad (5.4)$$

To solve the problem, the following mechanical and electrical boundary conditions are accepted:

(i) the boundary conditions in the form of a displacement field on the boundary S_u are given as

$$u_i|_{S_u} = u_i^0, \qquad (5.5)$$

(ii) the boundary conditions in the form of a vector of surface stresses P are given as

$$t = \sigma \, n|_{S_t} = P, \qquad (5.6)$$

(iii) the boundary conditions on the electrodes of the piezoelectric element $S_{\mathrm{E}} = \bigcup_k S_{E_k}$ are given as

$$\varphi|_{S_{E_k}} = \varphi_{0k}, \qquad (5.7)$$

(iv) the boundary conditions on non-electrode sections S_D, at the intersection of the corresponding areas $S = S_E \bigcup S_D$ are given as

$$D_n|_{S_D} = 0. \tag{5.8}$$

The damping coefficients are between the frequencies f_{r1} and f_{r2}. It is assumed that within the framework of the experiment, the change in the damping parameters α and β will be small. In the ANSYS FE software, the damping parameters are described in the form

$$\alpha = \frac{2\pi f_{r1} f_{r2}}{Q(f_{r1} + f_{r2})}, \quad \beta = \frac{1}{2\pi Q(f_{r1} + f_{r2})}. \tag{5.9}$$

From the experiment, the quality factor Q is found from the expression

$$Q = \frac{\omega}{\Delta\omega}, \tag{5.10}$$

where $\Delta\omega$ is the width of resonance curve. This is found at the corresponding resonance ω.

The set of Equations (5.1)–(5.8) is solved taking into account the initial boundary conditions for non-stationary problems [208]. The elements of solution for the set of Equations (5.1)–(5.8) within the framework of computer simulation can be implemented in the ANSYS software. Within the framework of the performed studies, direct calculations were carried out in the form of a modal analysis with obtaining natural modes of oscillations and frequency characteristics. By carrying out harmonic analysis, the electric potentials on the electrodes was calculated relative to the zero potential on certain surfaces. The displacement of the PEG pinching by 0.01 mm was taken as the perturbation parameter.

5.5. Model Studies of Cantilever-type PEG with Proof Mass and Piezoactive Elements

5.5.1. *Description of model parameters*

The cantilever-type piezoelectric transducer of mechanical energy into electrical contains a cantilever beam made of an elastic material on which piezoelectric elements are glued on one side (unimorph) or on both sides (bimorph). One end of the proof mass is disposed. Moreover, four piezoelectric elements, two above and two below relative to the beam, having either opposite or the same directions of the polarization vector, are installed at the base. The planes of their electrodes are pressed with the help of the base elements to the conductive layers of thin metallic elastic pads on one side (see Figure 5.9).

Thin symmetrical piezoelectric elements (PEs) are polarized in thickness and glued to the cantilever. The dimension characteristics of PEG elements are shown in Table 5.2, the mechanical properties of materials are present in Tables 5.3–5.5. Description of material parameters is given in Ref. [300]. The electrical and

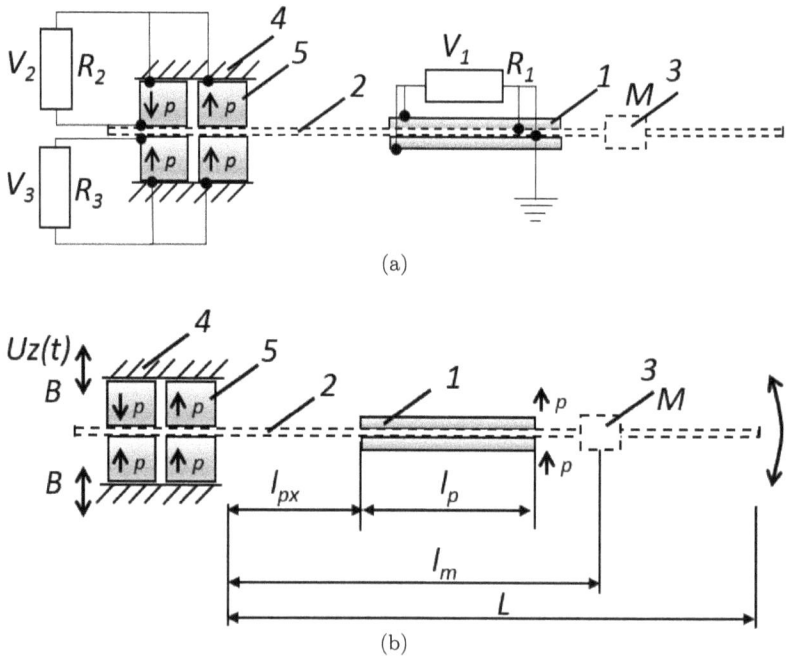

(a)

(b)

Fig. 5.9. (a) Representations of electrical scheme of PEG under active load with shown electrical parameters and (b) structural scheme of PEG with proof mass M and shown geometrical parameters ($U_z(t)$ is the oscillation law): 1 — piezoelectric element; 2 — substrate; 3 — proof mass; 4 — place of PEG fixing (B — movable base); 5 — piezoelectric cylindrical element.

Table 5.2. Dimension characteristics of PEG elements.

Piezoelectric elements			Piezoelectric cylinders	
l_p, mm	b_p, mm	h_p, mm	R, mm	H, mm
50	10	0.45	10	10

Substrate			Proof mass	
l, mm	b, mm	h, mm	M, g	l_m, mm
160	13.2	1.5	17.6	65–150

Table 5.3. Mechanical properties of structural materials.

Elements of PEG	Material	ρ, kg/m^3	$E \times 10^{10}$, Pa	ν
Base	Duraluminum	2800	0.33×10^{11}	0.33
Substrate	Duraluminum	2800	0.33×10^{11}	0.33
Proof mass	Steel	7700	21	0.33
Piezoelements	PCR-7M	7280	–	0.33
Piezocylinders	PZT-19	7280	–	0.33

Table 5.4. Elastic moduli C_{pq}^E (in 10^{10} Pa), piezoelectric coefficients e_{kl} (in C/m^2) and relative permittivity $\varepsilon_{kk}^\xi/\varepsilon_0$ of piezoceramics (at room temperature).

PEs type	C_{11}^E	C_{12}^E	C_{13}^E	C_{33}^E	C_{44}^E	e_{31}	e_{33}	e_{15}	$\dfrac{\varepsilon_{11}^\zeta}{\varepsilon_0}$	$\dfrac{\varepsilon_{33}^\zeta}{\varepsilon_0}$
PZT-19	10.9	6.1	5.4	9.3	2.4	−4.9	14.9	10.6	820	840
PCR-7M	13.3	9.2	9.1	12.5	2.28	−9.5	31.1	20.0	1980	1810

Table 5.5. Elastic compliance S_{pq}^E (in 10^{-12} Pa), piezoelectric moduli d_{fp} (in pC/H) and relative permittivity $\varepsilon_{kk}^\sigma/\varepsilon_0$ of piezoceramics (at room temperature).

PEs type	S_{11}^E	S_{12}^E	S_{13}^E	S_{33}^E	S_{44}^E	d_{31}	d_{33}	d_{15}	$\dfrac{\varepsilon_{11}^\sigma}{\varepsilon_0}$	$\dfrac{\varepsilon_{33}^\sigma}{\varepsilon_0}$
PZT-19	15.1	−5.76	−5.41	17.0	41.7	−126	307	442	1350	1500
PCR-7M	17.5	−6.7	−7.9	19.6	43.8	−350	760	880	3990	5000

structural schemes of the PEG are shown in Figure 5.9. The value of the proof mass varied from 3 to 25 g ($M = 17.6$ g was used in modeling).

5.5.2. *Computer simulation of PEG*

Figure 5.10 presents three-dimensional model of PEG with a proof mass, which can be displaced with respect to vertical y-axis (asymmetric case) and also changes its disposition along x-axis in different variants of calculation.

Three-dimensional finite element model of PEG in ANSYS is shown in Figure 5.11. Piezoelectric elements 1, 5 were modeled by using electroelastic finite element *SOLID5* with four degrees of freedom, electric potential *VOLT* and three components U_x, U_y, U_z of the displacement vector. Elastic finite element *SOLID186* was used for modeling substrate and proof mass. Moreover, specific finite element *CIRCU94* was used for modeling active load R.

5.5.3. *The results of modal calculation*

The results of modal calculations of the first 12 eigenmodes of PEG oscillations are given at the position of the proof mass, located near

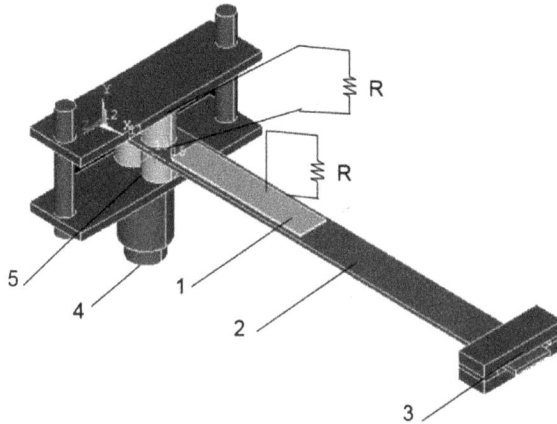

Fig. 5.10.   Three-dimensional finite element model of PEG: 1, 5 — piezoelectric elements; 2 — substrate; 3 — proof mass; 4 — place of generator fixing.

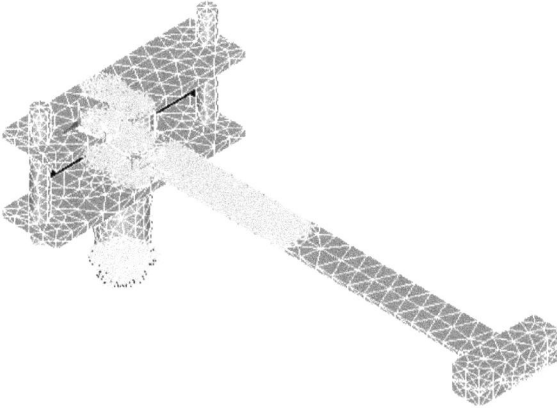

Fig. 5.11. Finite element model completed in ANSYS software.

the piezoelectric plates (see Figure 5.12). An analysis of the self-oscillation modes shows that 1, 2, 5, 7 and 9 forms of oscillations are transverse modes in the direction of the x-axis ($\boldsymbol{TrX}$). The modes of oscillations 3 and 6 are transverse with respect to the z-axis ($\boldsymbol{TrZ}$). The modes of oscillations 4 and 10 are torsional with respect to the x-axis ($\boldsymbol{TorX}$). The mode of oscillation 11 corresponds to the oscillation of base ($\boldsymbol{OscB}$) and the mode of oscillation 12 is a mixed mode ($\boldsymbol{OscB + TrZ}$).

The results of modal calculations of the natural frequencies and modes of PEG oscillations are presented in Figures 5.13 and 5.14 (first four eigenmodes are shown). The modes of oscillations depend on the position of the proof mass and change of the shape of oscillations. In this case, oscillation modes 2, 3, 4 with a smooth frequency displacement change the shape of the oscillations depending on the position of the proof mass. So, for example, 2 and 3 modes of the PEG oscillations at the position of the mass $L_m = 100$ change the shapes of each other: the bending shape of the oscillations relative to the x-axis TrX to the bending shape of the oscillations relative to the z-axis, TrZ. The bending shape 3 (TrX) replaces the torsional shape 4 (TorX) at the disposition of the proof mass $L_m = 117$.

Thus, mode Nos. 2, 3 and 4 can replace each other and be transitional at certain points of the location of the proof mass.

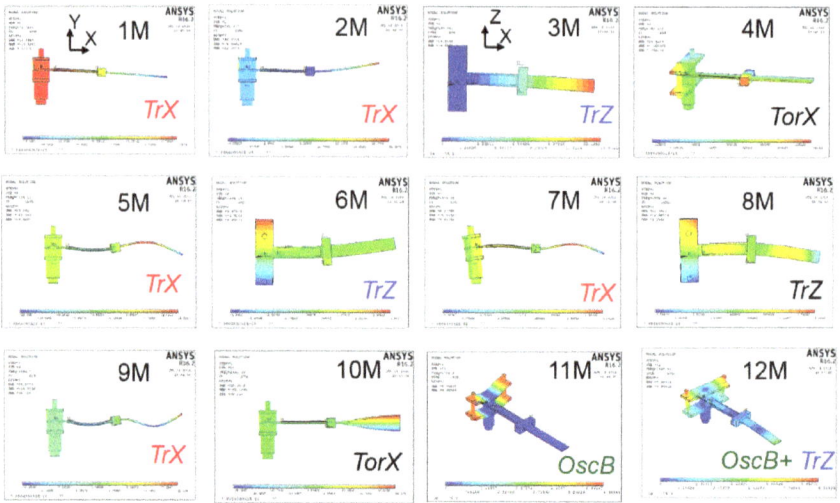

Fig. 5.12. Results of modal calculation for first 12 eigenmodes of PEG oscillations.

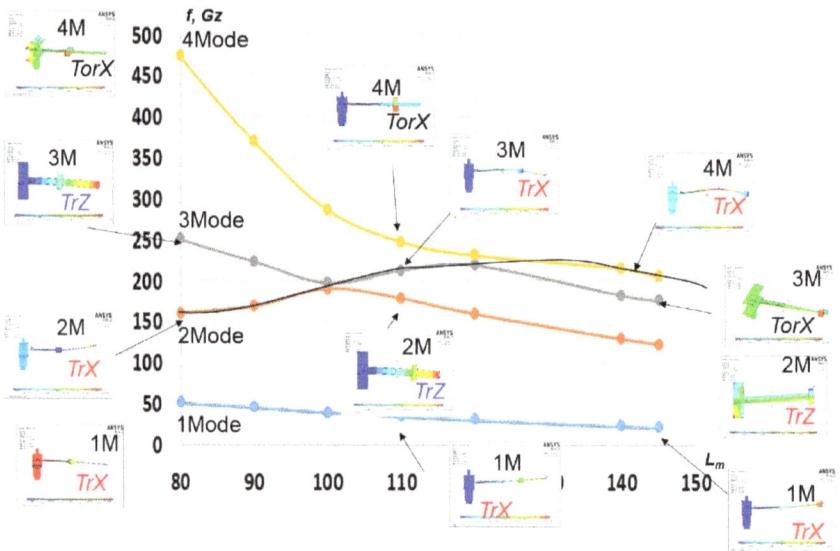

Fig. 5.13. Results of the modal calculation of first four eigenmodes.

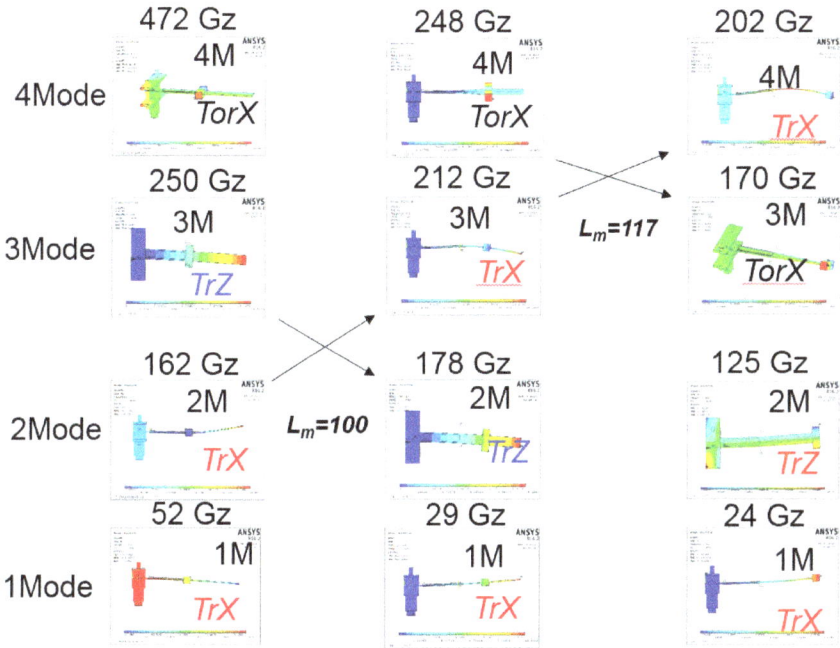

Fig. 5.14. Results of finite element modeling.

At the same time, the natural oscillation frequencies and the output voltage on the electrodes change, which must be taken into account in further design.

5.6. FE Simulation of PEG Output Characteristics at Asymmetrical Vertical Location of Proof Mass

5.6.1. *Three-dimensional model*

The considered full-scale FE model of PEG has a cantilever structure with a bimorph of thin symmetric piezoactive layers, polarized in thickness and glued on an elastic plate. The geometrical dimensions of PEG are shown in Figure 5.15(a): the substrate has dimensions $l \times b \times h = 150 \times 9.8 \times 1\,\text{mm}^3$, piezoelectric elements consist of two identical piezoelectric plates, polarized in thickness with dimensions $l_p \times b_p \times h_p = 54 \times 6 \times 0.5\,\text{mm}^3$. The center of proof

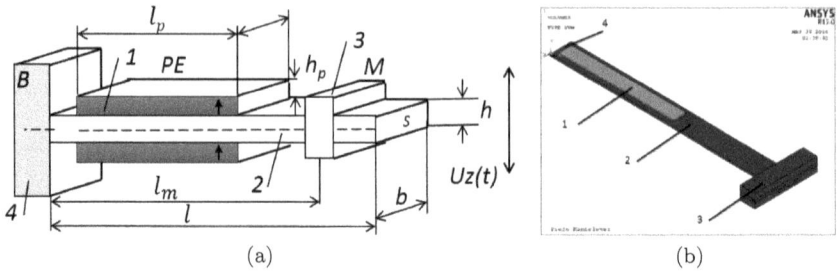

Fig. 5.15. Representations of (a) structural and (b) finite element PEG schemes: 1 — piezoelectric element, 2 — substrate, 3 — proof mass, 4 — clamp.

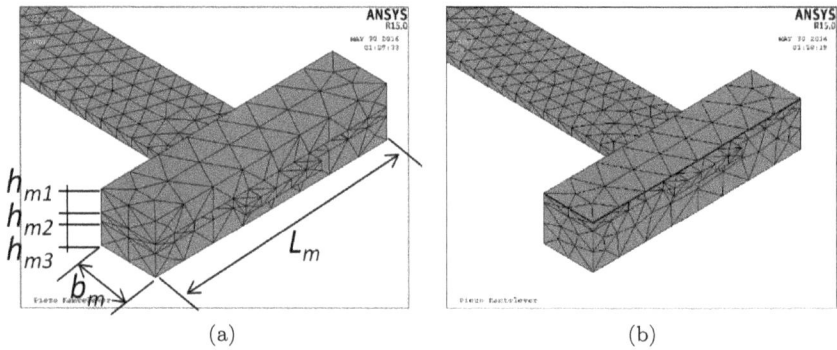

Fig. 5.16. Two types of the attachment of proof mass: (a) symmetrical height arrangement, (b) asymmetrical height arrangement.

mass is fixed at a distance l_m from the clamp of the cantilever (l_m can vary from 65 to 150 mm). The finite element partitioning and geometrical dimensions of the proof mass are shown in Figure 5.16 with symmetrical and asymmetrical heights of its location. The length L_m and width b_m of the inertial element (proof mass) are constant and equal to 8 mm and 22 mm, respectively. Thickness parameters h_{m1} and h_{m3} can vary from 0.1 to 5 mm, h_{m2} is the cantilever thickness equal to 1 mm. Value of the proof mass can vary from 5 to 25 grams. The material of piezoceramic plates is hot-pressed PCR-7M with parameters presented in Table 5.6. A PEG with three types of the substrate materials: duraluminum, steel and fiberglass, is considered. The density of duraluminum is more than the density

Table 5.6. The elastic moduli $C_{pq}^{E}(10^{10}\,\text{Pa})$, piezoelectric coefficients e_{kl} (in C/m^2) and relative permittivity $\varepsilon_{kk}^{\xi}/\varepsilon_0$ (at room temperature).

C_{11}^{E}	C_{12}^{E}	C_{13}^{E}	C_{33}^{E}	C_{44}^{E}	e_{31}	e_{33}	e_{15}	$\varepsilon_{11}^{\xi}/\varepsilon_0$	$\varepsilon_{33}^{\xi}/\varepsilon_0$
12.5	8.4	8.1	12.1	2.36	-9.0	28.3	17.9	1430	1350

Table 5.7. Mechanical properties of the structural materials.

Material	ρ, kg/m^3	$E \times 10^{10}$, Pa	ν
fiberglass	2500	7	0.25
steel	7800	21	0.3
duraluminum	2700	7.4	0.34

of fiberglass, but lower than the density of steel. The material of the proof mass is a fiberglass. The characteristics of the construction materials are shown in Table 5.7.

A three-dimensional finite element model of a PEG is shown in Figure 5.15(b). Piezoelectric medium 1 is modeled with electroelastic finite element *SOLID5* with four degrees of freedom: *VOLT* is the electric potential and U_X, U_Y, U_Z are the components of the displacement vector; substrate 2 and proof mass 3 are modeled by using elastic finite element *SOLID186*.

With the help of the developed FE models in ANSYS software based on the exact formulation of the problems (6.1) to (6.3), modal analysis and harmonic analysis were performed.

5.6.2. *Results of calculations*

At the first stage, the problem of comparison of the PEG output parameters in symmetric and asymmetric attachments of the proof mass at different electric resistive loads on each of the piezo-plates was solved. Location of the proof mass was $l_m = 65\,\text{mm}$. The output voltage of the piezoelectric plates 1 and 2 and output power of each plate for two cases of proof mass attachments are shown in Table 5.8.

Table 5.8. Output voltage and power of the each piezo-plate when location of proof mass was $l_m = 65\,\text{mm}$.

R_n, Ω	Asymmetric case				Symmetric case	
	U_1, V	U_2, V	$P_1, 10^{-6}\,\text{W}$	$P_2, 10^{-6}\,\text{W}$	$U_1 = U_2, \text{V}$	$P_1 = P_2, 10^{-6}\,\text{W}$
1000000	15.17	15.19	230.0	230.7	15.11	228.3
500000	14.19	14.17	402.8	401.5	14.12	398.7
250000	12.28	12.25	602.7	599.8	12.21	596.3
100000	8.18	8.17	669.6	667.3	8.15	664.2
50000	5.02	5.01	503.8	502.0	5.00	500.0
10000	1.17	1.17	137.1	136.7	1.17	136.0

Analysis of the results presented in Table 5.8 shows that the voltage between the piezoelectric plates with asymmetric and symmetric cases at various resistive loads does not differ much. The relative voltage deviation does not exceed 0.6%. As a result, asymmetrical arrangement of the proof mass has little effect on the output parameters of PEG voltage.

In the next step, the problem of calculating the dependence of the electric potential at the free electrode on the material properties and values of the substrate thickness was considered, while exposed to vibrations on non-resonance frequency of 10 Hz.

The dependencies of amplitude values of the electric potential on the free electrode on the material properties and a substrate thickness of 1.2 mm, while exposed to the vibrations on non-resonance frequency of 10 Hz, are shown in Figure 5.17. The length l of the substrate is 110 mm. The results for the cases of symmetrical and asymmetrical locations of a proof mass of 5 g are shown in Figures 5.17(a) and 5.17(b), respectively.

An analysis of the results obtained at non-resonance frequency 10 Hz, using substrates of different materials and attached proof mass, shows that the elastic characteristics of the substrate affect the output voltage PEG. By comparing the results with different mass locations, the ratio of the output voltage at $V(l_m = 65\,\text{mm})$ and $V(l_m = 150\,\text{mm})$ can be more than 4.

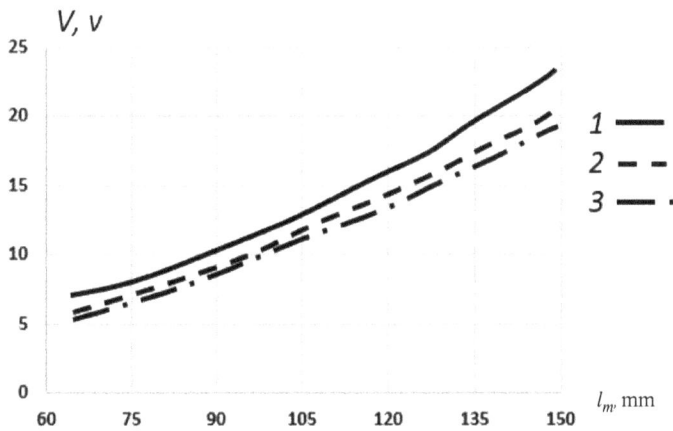

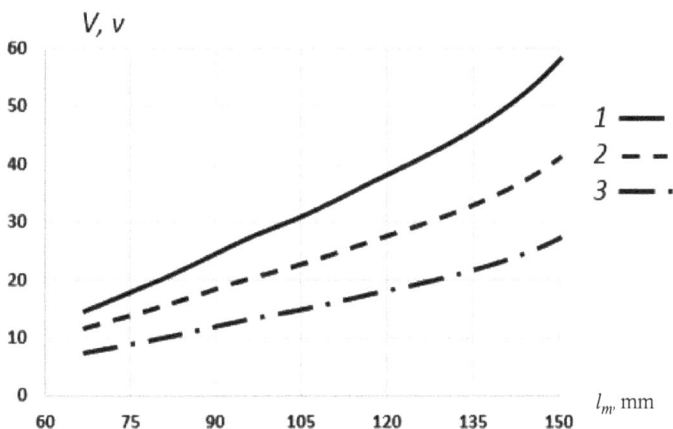

Fig. 5.17. (a) Symmetrical case and (b) asymmetrical case of 5 g proof mass location: 1 — fiberglass; 2 — duralumin; 3 — steel.

The results presented in Figure 5.17 show that the output voltage V at the non-resonance frequency 10 Hz is significantly less than the corresponding quantities in the case of resonance; however, these results allow to choose optimal materials and dimensions of the substrate and proof mass.

5.7. Modeling PEG under Pulsed Loading

5.7.1. *Formulation of problem and description of PEG model*

It is considered a cantilever-type PEG with active elements in the form of a bimorph and compressible piezocylinders under pulsed excitation at the base. This generator has plate-type piezoelectric elements experiencing bending deformations and cylindrical-type piezoelectric elements experiencing compressive strains. Non-stationary oscillations arise at the vibrational pulsed excitation of the PEG base.

Figure 5.18 shows the schematic representation of PEG, consisting of a circuit diagram of the device (Figure 5.18(a)) and PEG model (Figure 5.18(b)). The PEG includes the following elements: (1) generator mounting leg, (2) fixing frame, (3) substrate, (4) proof mass, (5, 6) plate piezoelectric elements (PEs), (7, 8) cylindrical PEs; $R_1 - R_4 = 1\,\mathrm{M\Omega}$ are the active electric loads of the corresponding PEs. The cylindrical PEs $(PE_1 - PE_4)$, polarized in thickness, have pairwise parallel electrical connections, respectively. The lower

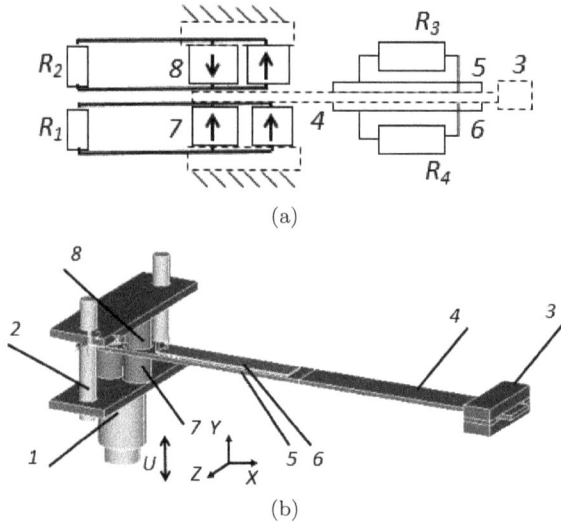

(a)

(b)

Fig. 5.18. Schematic representation of PEG: (a) circuit diagram; (b) model.

cylinders have codirectional polarizations, the upper cylinders have polarizations directed in opposite directions. Thin symmetrical PEs have a unidirectional polarization scheme relative to the y-axis and are fixed at the pinching of the console in a certain area. A detailed scheme of the model is presented in Ref. [376]. The value of proof mass in the calculations was equal to $M = 17.6$ g. The sizes of the piezoelectric plates were $l_p \times b_p \times h_p = 54 \times 6 \times 0.5$ mm^3, piezoelectric cylinders were $R \times H = 10 \times 10$ mm^2, the substrate was $l \times b \times h = 135 \times 13.2 \times 1.5$ mm^3 and the proof mass fixation was $l_m = 146.5$ mm.

5.7.2. *Results of numerical simulation*

PEG modeling was carried out in the ANSYS FE software. To build the model, the elements *SOLID5* and *SOLID45* of a three-dimensional configuration were used. The first type of elements is designed to model piezoelectric and elastic media, the second type of elements had a simplified stiffness matrix and simulated elastic media. In the process of building the model, 4100 nodes and more than 9200 elements were created. The solution of the problem in an unsteady statement was realized. The loading took place by simulating the movement of the PEG leg according to a linear law for simulating impact loading.

At the first stage, a modal analysis of the structure was performed. In the FE model, the displacements of nodes on the surface of the PEG legs in three directions were recorded. Four natural frequencies of PEG vibrations were obtained: $\omega_1 = 36.928$ Hz, $\omega_2 = 133.63$ Hz, $\omega_3 = 293.35$ Hz, $\omega_4 = 378.01$ Hz. The calculation results are presented in Figure 5.19.

At the second stage, harmonic analysis was performed for stationary PEG excitation by displacement relative to the y-axis, applied to nodes, located on the fixed plane at the generator base. The displacements of the nodes of this plane in the remaining directions were fixed. For analysis, we considered the frequency spectrum ω_i from 1 to 200 Hz. Figure 5.20 presents the dependences of the amplitude of the oscillations of the points P_1, P_2, P_3 (at left, mean

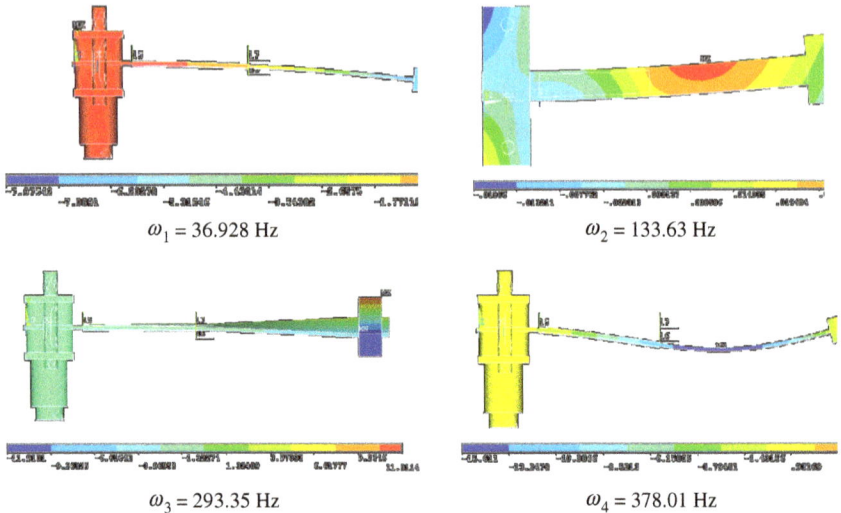

$\omega_1 = 36.928$ Hz

$\omega_2 = 133.63$ Hz

$\omega_3 = 293.35$ Hz

$\omega_4 = 378.01$ Hz

Fig. 5.19. First four natural frequencies and the corresponding waveforms of the PEG model.

and right points of the cantilever, respectively), located on the cantilever of the generator and dependences of the voltages across the electrodes of the piezoelectric elements of the generator $PE_1 - PE_4$ on the frequencies during harmonic analysis.

An analysis of the plots shows that in the frequency range from 1 to 200 Hz in the excited y-direction of the PEG base vibrations, the maximum output voltage was reached with resonance of the cantilever bending vibrations at a frequency of 33 Hz. In this case, the values of the voltage on the PE plates are equal, respectively: $V_1 = 0.0869$ V, $V_2 = 0.138775$ V, $V_3 = -4.17596$ V, $V_4 = 4.17596$ V. One of the electrodes, located on the cantilever, will have a minimum output voltage, thereby working in antiphase of antiresonance. The maximum oscillation parameters are achieved at the bimorph on the base at a frequency of 48 Hz. In this case, the output voltage parameters of the corresponding piezoelectric elements are: $V_1 = 0.0361$ V, $V_2 = 0.0769$ V, $V_3 = 4.65054$ V, $V_4 = -4.65054$ V.

Peak characteristics of the output voltage of cylindrical PEs are achieved at a resonance frequency of $\omega_1 = 36.928$ Hz. In this case,

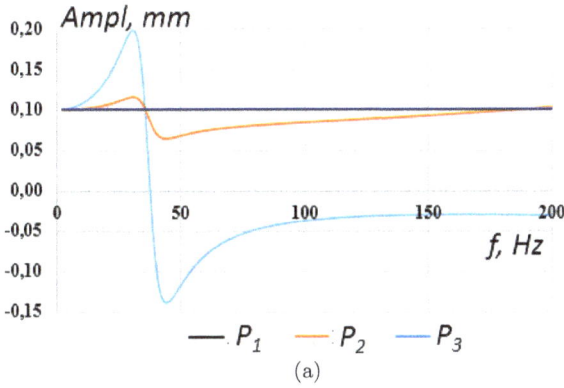

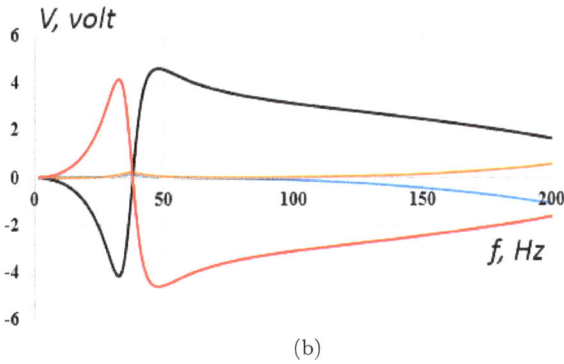

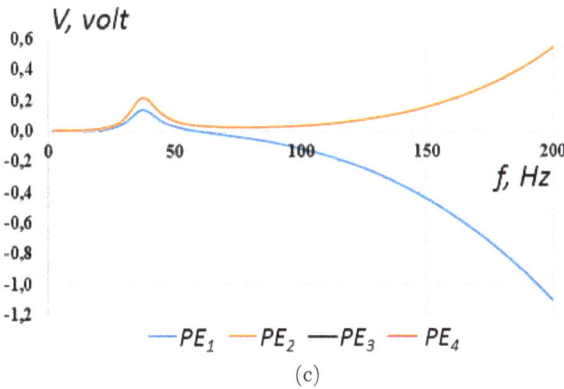

Fig. 5.20. (a) Dependences of the amplitude of oscillations at the points P_1, P_2, P_3, located on the generator cantilever, in the vertical y-direction; (b, c) dependences of the voltage on the electrodes of the piezoelectric elements of the generator $PE_1 - PE_4$, depending on the frequencies during harmonic analysis.

the parameters of the output voltage characteristics were: $V_1 = 0.1345734\,\text{V}, V_2 = 0.214175\,\text{V}, V_3 = -1.56457\,\text{V}, V_4 = 1.564575\text{V}$.

5.7.3. *Numerical simulation of PEG with active base*

The problem of pulsed excitation of a piezoelectric generator with active base was considered, too. As input parameters, the displacement field at the cantilever base as a result of short-term pulse impact was studied. A pulsed displacement of a certain amplitude $U = 0.1\,\text{mm}$ was applied to the point 1 of the PEG surface (see Figure 5.21(a)). The values of displacement application time T_{imp} were considered as $0.025\,\text{s}, 0.01355\,\text{s}, 0.01\,\text{s}$ and $2\,\text{s}$. This corresponded to the work of PEG at half-wave frequencies f_{imp} of pulsed action of $0.25\,\text{Hz}, 20\,\text{Hz}, 37\,\text{Hz}$ and $50\,\text{Hz}$. Unsteady oscillations of the generator were analyzed at above-mentioned points P_1, P_2, P_3 and the values of voltage on the plates of the generator electrodes were found for time $t = 5\,\text{s}$.

In ANSYS FE software, the attenuation coefficients α and β were introduced to describe the losses of mechanical energy. By assuming the same Q-factor at the first two f_{r1}, f_{r2} resonance frequencies, they were found through the Q-factor (in this study, the quality factor of all PEG materials was taken equal to $Q = 10$) as follows [30]:

$$\alpha = \frac{2\pi f_{r1} f_{r2}}{Q(f_{r1} + f_{r2})}, \quad \beta = \frac{1}{2\pi Q(f_{r1} + f_{r2})} \tag{5.11}$$

Due to pulsed loading, a wave of displacements of the cantilever points arose. In the process of solving the problem, displacements were analyzed at control points 1–3 of the cantilever. Figure 5.21 shows the transverse absolute displacements at points P_1, P_2, P_3 of the cantilever surface. Figure 5.22 demonstrates corresponding values of output voltage at excitation of the piezoelectric elements in the cantilever during pulsed loading for the considered cases.

The average value of the instantaneous power $\overline{P}$ over a period T was calculated by formula (5.12) and the work A, performed by corresponding piezoelectric elements for the period T, was calculated

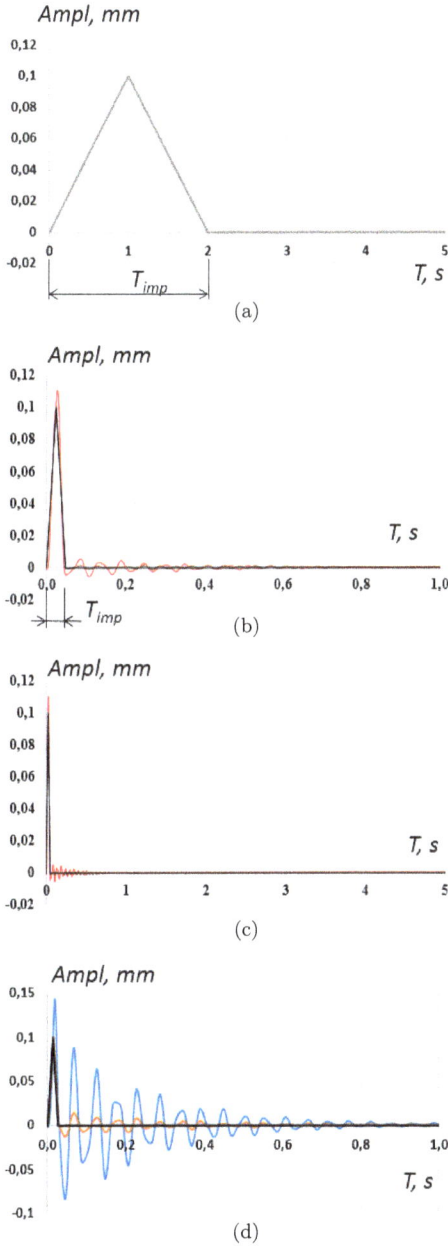

Fig. 5.21. Oscillation amplitudes of the cantilever points: ▬ P_1 ▬ P_2 ▬ P_3 upon pulsed excitation at the time T. The time of the load application: (a) $T_{\mathrm{imp}} = 2\,\mathrm{s}$; (b), (c) $T_{\mathrm{imp}} = 0.025\,\mathrm{s}$; (d), (e) $T_{\mathrm{imp}} = 0.01355\,\mathrm{s}$; (f), (g) $T_{\mathrm{imp}} = 0.01\,\mathrm{s}$.

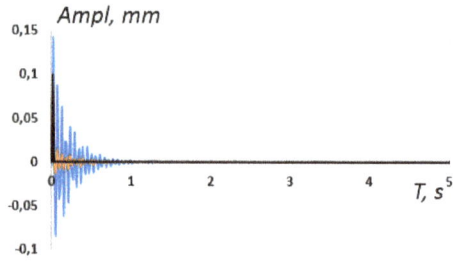

(e)

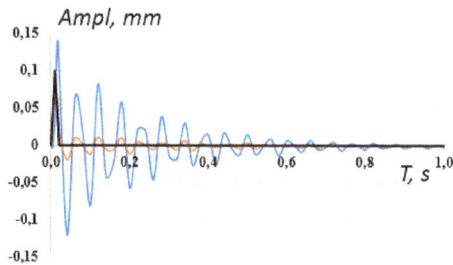

(f)

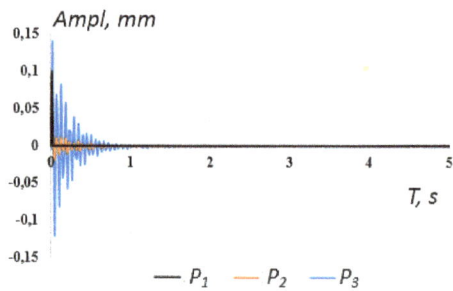

(g)

Fig. 5.21. (*Continued*)

by formula (5.13):

$$\overline{P} = \frac{1}{T} \int_0^T p(t)dt = \frac{1}{T} \int_0^T \frac{V(t)}{R} dt, \qquad (5.12)$$

$$A = \int_0^T p(t)dt. \qquad (5.13)$$

The calculated values of the average instantaneous power $\overline{P}$ and the work A, performed by various PEs over a period $T = 1\,\mathrm{s}$,

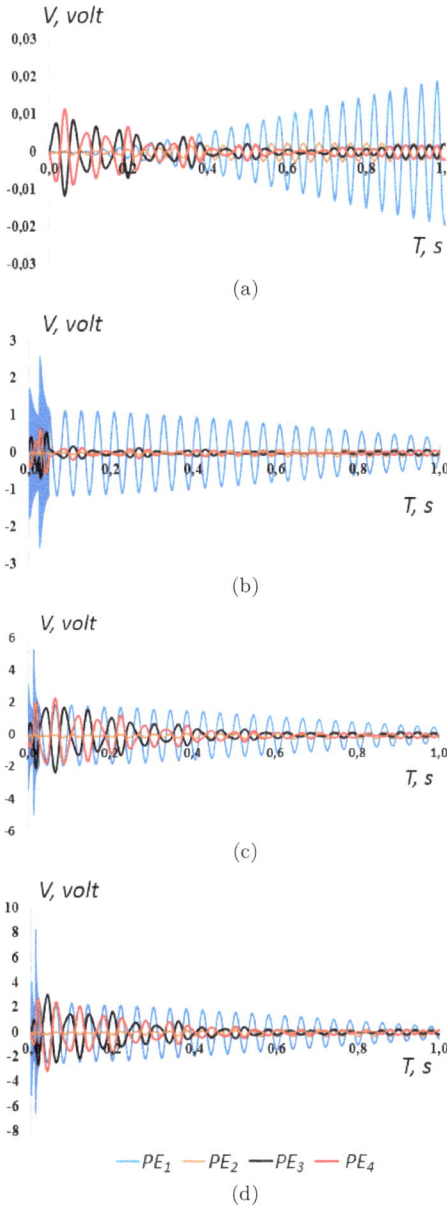

Fig. 5.22. Output voltage at excitation of the piezoelectric elements in the cantilever during pulsed loading for a time $T = 1\,\mathrm{s}(\mathrm{a,b,c,d})$; color curve corresponds to PEs; time of pulsed load application: (a) $T_{\mathrm{imp}} = 2\,\mathrm{s}$; (b) $T_{\mathrm{imp}} = 0.025\,\mathrm{s}$; (c) $T_{\mathrm{imp}} = 0.01355\,\mathrm{s}$; (d) $T_{\mathrm{imp}} = 0.01\,\mathrm{s}$.

Table 5.9. Calculated values of average instantaneous power $\overline{P}$ and work A performed by various PEs over a certain period $T = 1\,\text{s}$.

No.	Parameter	Piezoelectric element number			
		PE$_1$	PE$_2$	PE$_3$	PE$_4$
1	$\overline{P}$, μW	0.00006460	0.00000313	0.00000526	0.00000526
	A, μVA	0.00006460	0.00000313	0.00000526	0.00000526
2	$\overline{P}$, μW	0.73800	0.00603	0.00430	0.00430
	A, μVA	0.73800	0.00603	0.00430	0.00430
3	$\overline{P}$, μW	1.7100	0.0218	0.1900	0.1900
	A, μVA	1.7100	0.0218	0.1900	0.1900
4	$\overline{P}$, μW	2.550	0.041	0.288	0.288
	A, μVA	2.550	0.041	0.288	0.288

are present in Table 5.9. The calculations show that the process of attenuation occurs within 5 s for case 1 at $T_{\text{imp1}} = 2\,\text{s}$. Due to this, a comparative period of 5 s was chosen for average power and performed work. An analysis of the average power $\overline{P}$ and the performed work for various PEs shows that highest values are given by a pair of cylindrical PE$_1$, for all variations of the PEG excitation time. Owing to the neglecting of the scale factor for the cylindrical piezoelectric element and proof mass (inertial load for PE at the base of construction), additional calculations are required to establish the influence of these factors. For example, under the pulsed impact during $T_{\text{imp}} = 2\,\text{s}$, the minimum values of the average power were obtained, although in the case of the pulsed loading during $T_{\text{imp}} = 0.01\,\text{s}$, the values of both the average power and performed work were maximum for all PEs.

Thus, the performed modal, harmonic and non-stationary analysis of PEG leads to the following results. The harmonic analysis shows that the maximum, output parameters are achieved in the bimorph piezoelements at a frequency of 48 Hz, and in the cylindrical piezoelements of the PEG base at a frequency of 33 Hz. For non-stationary calculation, it was shown that the maximum average power value can attain 2.550 μW for cylindrical PEs with unidirectional polarization at a resistance of 1 MΩ, and 0.041 μW in the case of mixed polarization. Moreover, for plate PEs, located on the cantilever, the power is equal to $0.288\,\mu\text{W}$. These power values were obtained during shock excitation of the generator base during $T_{\text{imp1}} = 0.01$ s and a displacement amplitude of $0.1\,\text{mm}$.

5.8. Cantilever-type PEG with Porous Piezoceramic Elements

Let us consider cantilever-type PEGs, presented in Figures 5.10 and 5.11, with porous piezoceramic elements. Effective properties of cylindrical and plate PEs with different porosities are defined on the basis of Ref. [290] and correspond to piezoceramic material PZT-4. The results of calculating the effective moduli of this piezo-composite with a 3-0 connectivity type for various porosities from 10% to 80% are shown in Table 5.10 [289]. Ceramic PZT-4 was chosen as the main material (first column of Table 5.10).

Figure 5.23 shows the dependencies of the effective elastic stiffness moduli and density of the piezoceramic material on the porosity.

Table 5.10. Effective material properties of the porous composite: ρ is the density, $c_{ij}^{E\text{eff}}$ are the elastic stiffness moduli, e_{ij}^{eff} are the piezoelectric moduli, $\varepsilon_{ij}^{\text{eff}}$ are the dielectric permittivity moduli.

% of porosity	0	10	20	30	40	50	60	70	80
$\rho,\text{kg/m}^3$	7500	6750	6000	5250	4500	3750	3000	2250	1500
$c_{11}^{E\text{eff}},10^{10},\text{N/m}^2$	13.9	11.56	9.25	6.85	5.05	3.34	2.07	1.26	0.68
$c_{12}^{E\text{eff}},10^{10},\text{N/m}^2$	7.78	6.15	4.66	3.14	2.10	1.16	0.62	0.28	0.13
$c_{13}^{E\text{eff}},10^{10},\text{N/m}^2$	7.43	5.82	4.25	2.82	1.87	1.06	0.52	0.24	0.1
$c_{33}^{E\text{eff}},10^{10},\text{N/m}^2$	11.5	9.53	7.23	5.42	3.91	2.72	1.63	0.91	0.47
$c_{44}^{E\text{eff}},10^{10},\text{N/m}^2$	2.56	2.23	1.83	1.44	1.10	0.74	0.44	0.23	0.1
$e_{33}^{\text{eff}},\text{C/m}^2$	15.1	13.38	11.37	9.59	7.68	5.93	3.93	2.30	1.25
$e_{31}^{\text{eff}},\text{C/m}^2$	-5.2	-4.23	-3.14	-2.07	-1.32	-0.75	-0.43	-0.21	-0.1
$e_{31}^{\text{eff}},\text{C/m}^2$	12.7	10.96	8.96	6.91	5.00	3.30	1.95	1.00	0.44
$\varepsilon_{11}^{\text{eff}}/\varepsilon_0$	730	663	582	509	439	349	263	191	122
$\varepsilon_{33}^{\text{eff}}/\varepsilon_0$	635	567	492	413	345	270	199	130	75

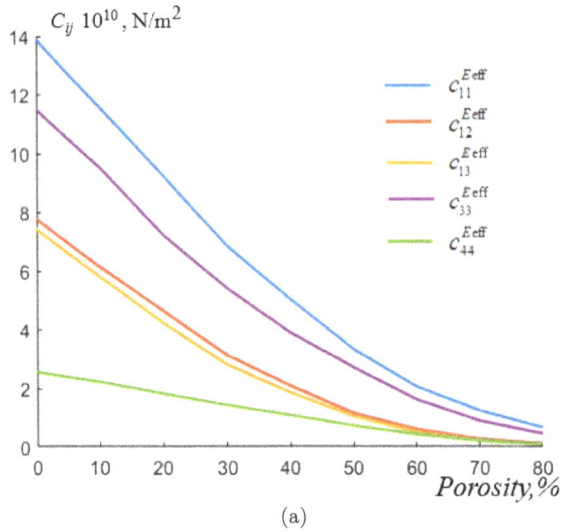

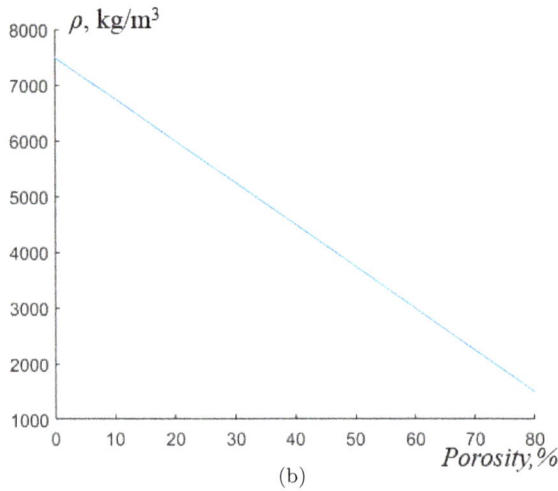

Fig. 5.23. (a) Dependences of elastic stiffness moduli, and (b) density of piezoelectric ceramics on different values of porosity.

5.8.1. *FE modeling of natural frequencies*

The first part of the numerical experiment consisted in definition of the PEG natural frequencies and natural forms. The simulation was carried out with an active load of piezoelectric elements, $R = 1000\,\Omega$.

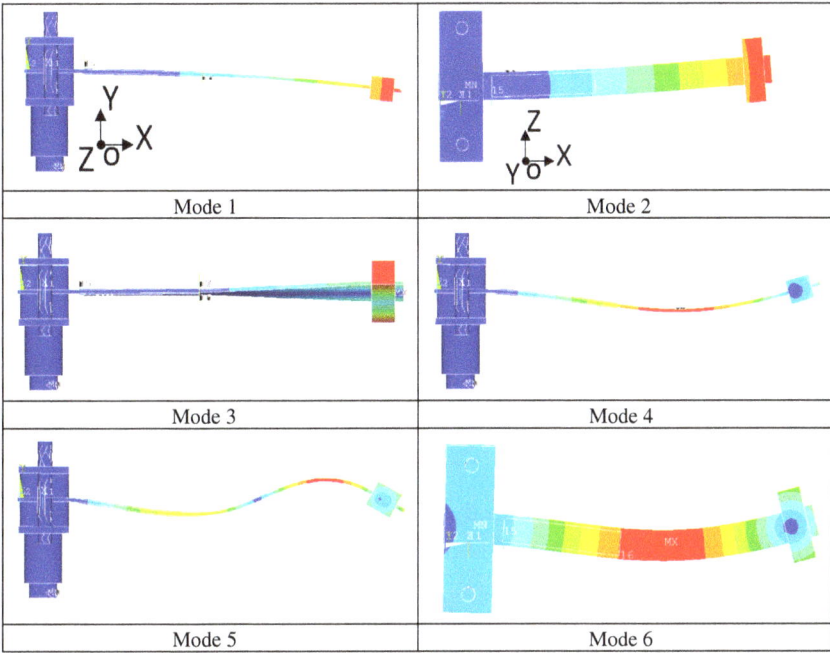

Fig. 5.24. Eigenmodes of PEGs.

A proof mass of 17.6 g was located at a distance equal to 146.5 mm. Figure 5.24 shows the first 6 modes of natural oscillations calculated in the ANSYS software. A model with 0% porosity was considered as a basic model (for comparison). The first oscillation mode had a frequency of 37.8 Hz, and was a flexural mode about the x-axis. First four oscillation modes changed in the range up to 382.5 Hz. Oscillation modes 5 and 6 have frequencies 1058.7 and 1318.7 Hz, respectively. The most effective are the oscillation modes 1 (ω_1 = 37.8 Hz), 4 (ω_4 = 382.5 Hz) and 5 (ω_5 = 1058.7 Hz), being bending modes with respect to the x-axis. In the Oxy-plane, at the lowest stiffness of the PEG cantilever, they form the bending oscillation modes 1, 2 and 3. The oscillation modes 2 and 6 are flexural with respect to the x-axis in the plane of the highest cantilever rigidity. They form in this plane the vibration modes 1 and 2, respectively. The natural mode 3 of PEG vibration is torsional about the x-axis.

Table 5.11. Dependence of the natural frequencies of the generator on the porosity of the piezoelectric elements.

Mode No.	Porosity of ceramics (%)								
	0	10	20	30	40	50	60	70	80
	Natural frequencies, ω_i								
1	37.8	37.2	36.5	35.7	34.9	33.9	32.9	32.1	31.4
2	136.3	134.4	131.8	128.8	125.8	122.1	117.7	112.7	104.9
3	295.4	294.3	292.9	291.1	289.4	287.4	285.0	283.0	280.7
4	382.5	382.1	380.9	378.9	376.9	374.2	371.1	368.2	364.6
5	1058.7	1063.1	1066.3	1068.2	1070.4	1071.3	1071.4	1072.4	1071.2
6	1318.7	1321.3	1319.7	1315.7	1308.6	1294.6	1267.5	1223.6	1152.2

5.8.2. *Harmonic analysis*

The results of harmonic analysis of oscillations were obtained by using ANSYS FE software. The calculation took into account the magnitude of the acceleration equal to $0.01\,\mathrm{m/s^2}$ and applied to the entire construction in the frequency range from 1 to $600\,\mathrm{Hz}$. Table 5.11 shows the results of calculations of the dependence of the natural frequencies of the generator on the porosity of the piezoelectric elements. The graphical dependences of the absolute and relative values of the natural frequencies for various oscillation modes are present in Figure 5.25(a) and 5.25(b). Analysis shows that the frequency of mode 1 decreases more than 15% with growth of porosity from 0% to 80%. The frequency of the oscillation mode 2 drops down to 23% at a porosity of 80%. The frequencies mode 3 and the subsequent vibration modes vary within 5%. The most effective vibration mode is the first flexural vibration frequency.

5.9. Axial-Type PEG with Constructive L-shaped Elements

5.9.1. *Model parameters and electrical scheme*

New axial-type PEG for transforming mechanical energy into electrical one consists of a base plate in the shape of a rigid beam construction with glued unimorph or bimorph PEs. One end of the beam structure is bolted to the base. At the other end of the base,

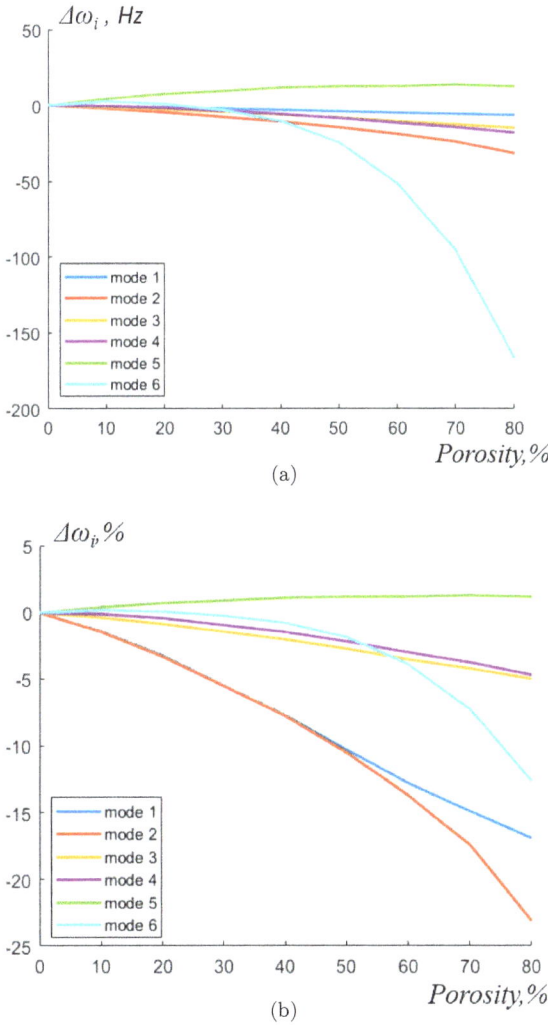

Fig. 5.25. (a) Absolute and (b) relative dependences of the natural frequencies of the generator on the porosity of bimorph and piezocylindrical elements ($\Delta\omega_{i,P} = \frac{\omega_{i,P} - \omega_{i,P=0}}{\omega_{i,P=0}}$).

cylindrical PEs are fixed coaxially with the base bar by means of an L-shaped bar. A proof mass is disposed between the edge of the base, fixed with a bolt, and the piezoelectric elements. PEs can have polarization vectors in various directions. The planes of cylindrical

PEs with electrodes are pressed by means of L-shaped elements to the conductive layers of thin elastic spacers, metalized on one side. The PEG is located on a specialized platform (see Figure 5.26), which allows one to fix the PEG elements rigidly.

Thin symmetric piezoelectric elements from ceramic PZT-19 are polarized in thickness. They are glued to the base console and are arranged in a row (see Figure 5.27). Characteristics of the dimensions of the PEG elements are presented in Table 5.12.

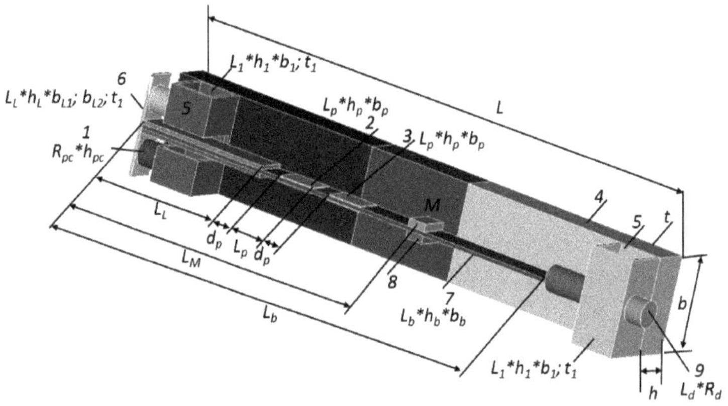

Fig. 5.26. Representation of the structural scheme of axial-type PEG with proof mass: 1 — cylindrical PE; 2, 3 — plate PEs; 4 — rigid platform of PEG; 5 — PEG fixing supports; 6 — L-clamping bar; 7 — base console; 8 — proof mass; 9 — screw.

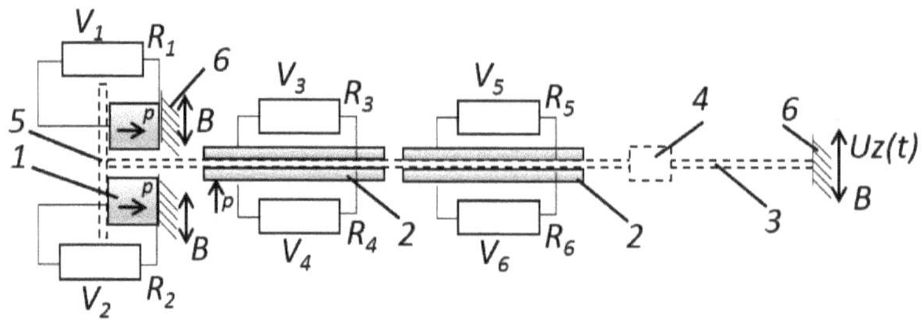

Fig. 5.27. Representation of the Electrical scheme of axial-type PEG under active electric load: 1 — cylindric PE; 2 — bimorph PE; 3 — substrate; 4 — proof mass; 5 — clamping L-shaped bar of cylindric PEs fixing; 6 — place of PEG fixing (B is the movable base); p is the PEs polarization direction.

Table 5.12. Geometric characteristics of the dimensions of PEG elements.

Name		Geometric parameters, mm							
Piezo cylinder	R_{pc}	14	h_{pc}	12.5					
Piezo bimorph	L_p	20	b_p	15	h_p	0.5			
Rigid platform	L	300	b	50	h	25	t	2	
Fixing supports	L_1	50	b_1	25	h_1	25	t	2	
L-clamping bar	L_L	74	b_{L1}	25	b_{L2}	24			
	t_1	15	h_L	2			d_p	10	
Base console	L_b	260	b_m	14	h_m	1.5			
Mass	L_m	16	b_m	10	h_m	4			
Screw	L_d	44	R_d	13					

An electrical scheme of a PEG connection with a resistive load is shown in Figure 5.27. The resistive load is supplied to each PE individually. Voltage is found at the contact points of the resistor. In numerical simulation, voltage is calculated as the difference of its amplitudes at the nodes of the FE element presented with an option indicating the type of resistor.

5.9.2. *FE modeling*

Figure 5.28 shows the FE model of the PEG in ANSYS software. During the PEG modeling, elements of the *SOLID92* type with a tetraidal structure were used for partition of the model. By modeling the thin walls of the planks, it was assumed not to consider deformations over the thickness of the construction. As a result, the value of the smallest edge of the FE in the form of a tetrahedron was taken equal to the wall thickness of the structural element. Modeling the PEs was carried out by using FE of the *SOLID5* type. Modeling of elements in the form of active resistance was carried out using the FE *CIRCU94* with the resistor option.

The cylindrical PE of a three-dimensional structure was divided by an FE mesh with an edge size multipled by 0.15 of the piezocylinder height. For plate PE, the size of the partitioning of the FE mesh by thickness was a multiple of the value equal to the thickness of the PE. It had a shape of triangular prisms. The direction of polarity in

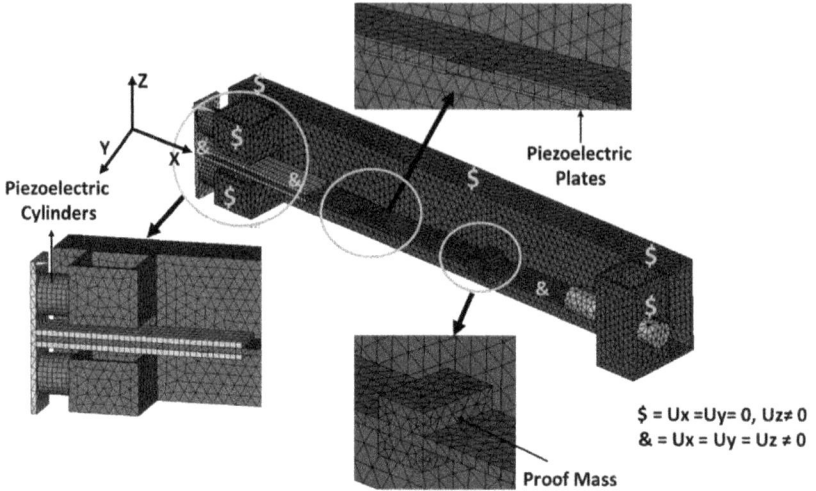

Fig. 5.28. FE model of PEG of axial type and with boundary conditions.

a cylindrical PE was taken along the main axis of the PE, coaxially with the PEG; it was along the thickness in a plate-shaped PE. The polarization vector in the bimorph for the upper and lower PEs was directed along the normal to the surface. By partitioning the model, the number of FE elements was more than 33,000 and the number of nodes was more than 62,500.

5.9.3. *Modal analysis*

At the first stage, modal analysis of PEG was performed. Figure 5.29 presents first 10 oscillation modes of the model with a proof mass of 0.01 g (conditionally, the minimum value of the mass in the simulation). It was assumed that the configuration of proof mass had an insignificant effect on the modal characteristics of the PEG. The model had a first design frequency of 259 Hz. The oscillation mode for the mode 1 had a prevailing bending character for the base plank of the generator. Oscillation modes 1, 4, 8 corresponded to (1, 2, 3) flexural oscillation modes of the PEG base plank in the vertical direction. It was assumed that bending strains of the base plank in the smallest plane of rigidity (in the vertical direction) excited the highest output voltage in all PEs.

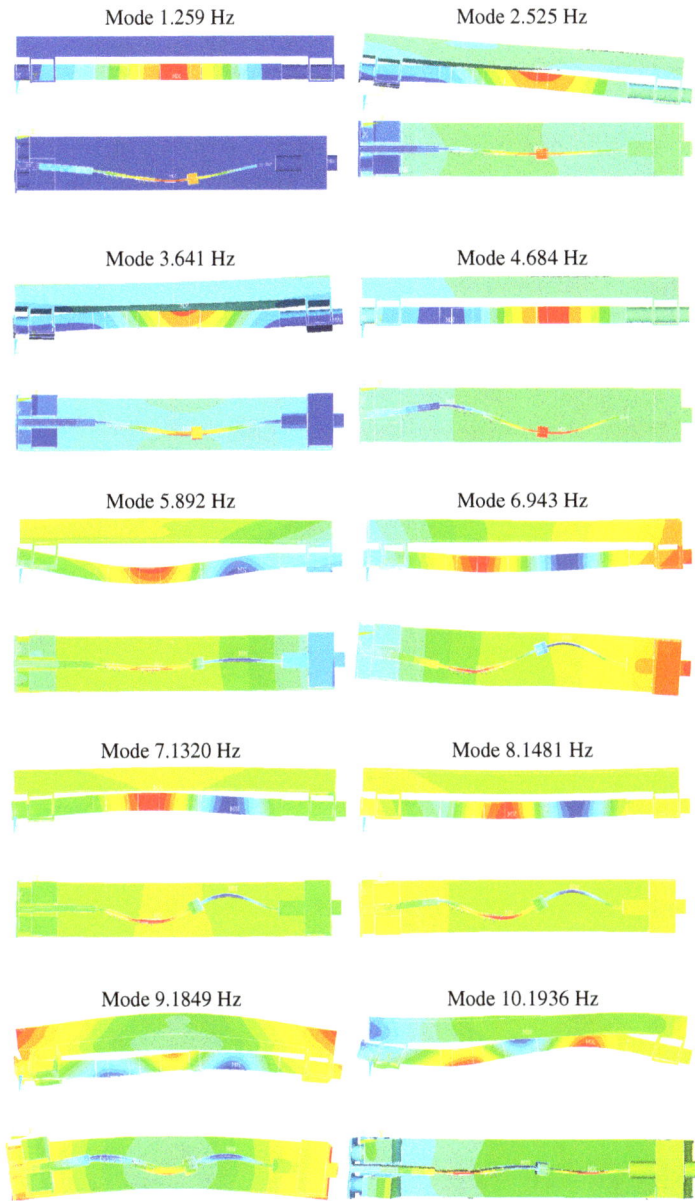

Fig. 5.29. Natural frequencies and oscillation modes for the first 10 modes of the model with proof mass. The conditional color gradient of the structure deformation in the vertical direction is given.

5.9.4. *Harmonic analysis*

Investigations of oscillations of axial-type PEG under harmonic action have been carried out. The harmonic action was calculated under the action of a uniformly applied acceleration of 10 m/s² for all structural units at the corresponding harmonics. For the oscillation modes 1 and 2, their dependences on the value and location of the proof mass were defined. The corresponding plots (see Figure 5.30)

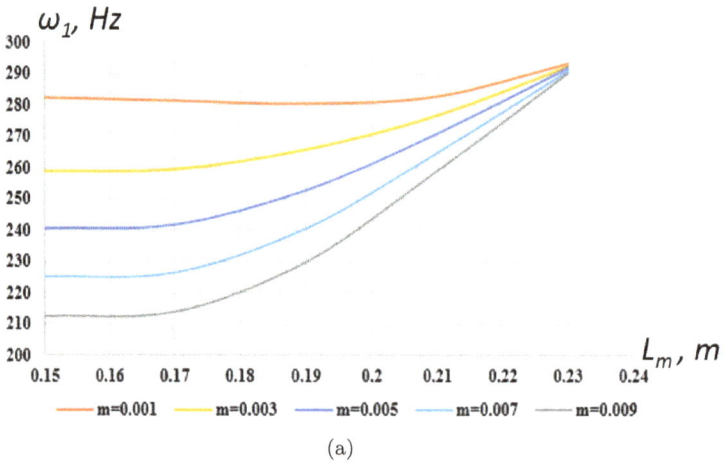

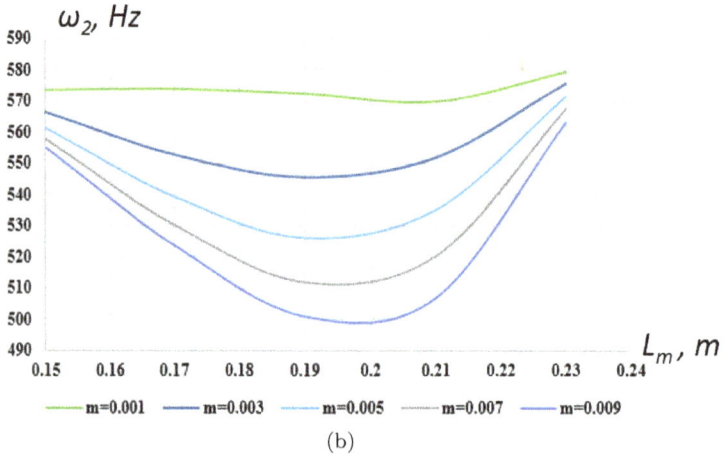

(a)

(b)

Fig. 5.30. Dependences of 1st and 2nd natural frequencies of PEG on the magnitude m (in kg) and location L_m of proof mass.

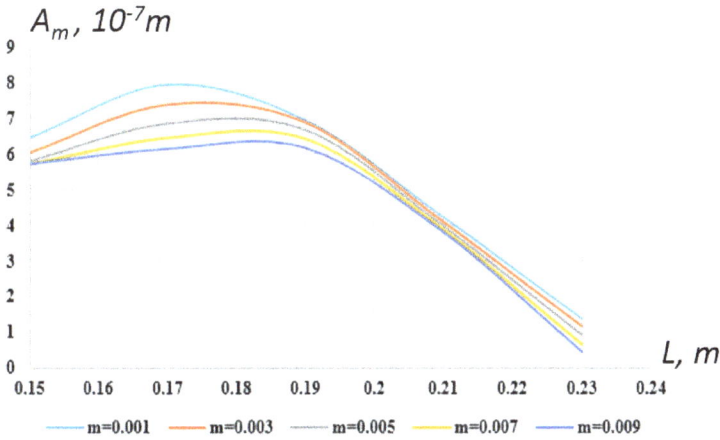

Fig. 5.31. Amplitude of steady-state oscillations of proof mass in dependence on rigid platform L of PEG at 1st natural frequency.

show that the lowest frequency values are in the range of 212–220 Hz for the oscillation mode 1 and values of 501–510 Hz can be achieved at the central disposition of the proof mass and its maximum value of 9 g. The oscillation amplitude at the point of the proof mass attachment is maximum (see, Figure 5.31). In Figure 5.32, the calculated parameters of the dependence of the output voltage at the electrodes of piezocylinders (U_1) and piezoelectric bimorphs (U_2 and U_3), respectively, at their calculated arrangement from left to right (see Figure 5.26). In the calculations, the value of the active electric load of the corresponding PEs was taken equal to 1000 Ω for the oscillation mode 1.

5.10. Conclusions

This chapter has presented in a more detailed consideration some piezoelectric power generation devices, discussed in Chapter 4. In particular, experimental and finite element models were devoted to cantilever-type and axial-type piezoelectric generators with active elements. Laboratory test set-ups were described in necessary details. The obtained results covered different structural and electrical schemes of the PEGs with proof mass (at its symmetrical and

(a)

(b)

(c)

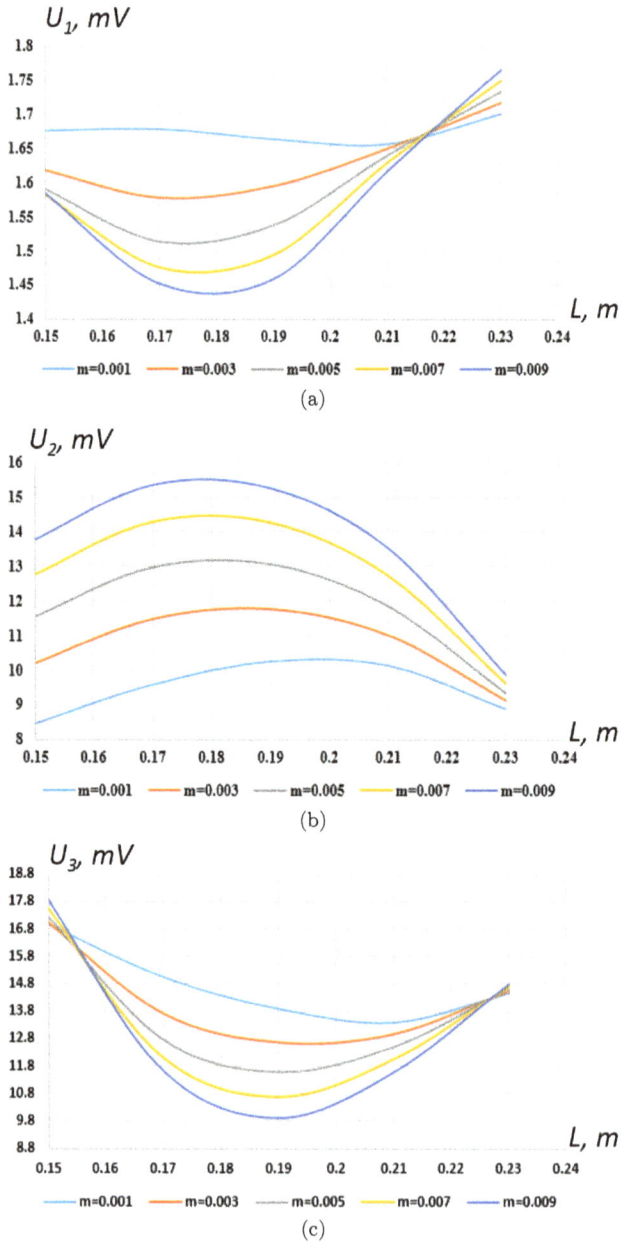

Fig. 5.32. Dependence of the output voltage on the electrodes of the corresponding PEs on magnitude of proof mass m (in kg) and rigid platform L of PEG for vibration mode 1 at active electric load of 1000 Ω.

asymmetrical locations), piezoelectric elements of bimorph and cylindrical types, and excitation loads (harmonic and pulse). Based on finite element modeling, the results of modal and harmonical analysis have been obtained. Resonance frequencies and output characteristics (voltage and power) were calculated taking into account geometrical and physico-mechanical properties of proof mass and other constructive elements. This information is necessary for designing optimal PEG constructions with high piezoelectric energy harvesting characteristics.

Chapter 6

Modeling of Oscillations of Devices Based on Dielectrics under the Influence of Fields of Various Nature: The Current State of the Art

6.1. Introduction

Dielectrics represent a wide class of materials with different properties. The main property of a dielectric material is the ability to polarize in an external electric field. In addition, in various classes of dielectric materials, effects were discovered that are described by the relationship between physical fields of different nature. The study of these multiphysical effects laid the foundation for separate directions in science and new classes of devices in applied instrumentation. Such effects include piezoelectric, pyroelectric, electrocaloric and magnetostrictive.

This chapter reviews the work done over the last 5 years on modeling vibrations of devices that exhibit one or more multiphysical effects. The main emphasis is placed on the interaction of the mechanical field with the electric, temperature and magnetic fields. Therefore, such effects as piezoelectric effect, pyroelectric effect and piezomagnetic effect are considered. Moreover, studies on high-order effects associated with the interaction of deformation and polarization gradients, namely, the flexoelectric and flexomagnetic effects, are considered. These effects can manifest themselves in a narrower class of dielectrics called piezoactive materials. Such materials have found their way into various devices such as sensors, actuators and generators. Modeling vibrations of devices based on piezoactive materials is an important task in various fields, including medical technology, aviation and automotive industries.

There are several main methods for modeling devices based on piezoactive materials: analytical, semi-analytical and finite element analysis.

The analytical method is used for solving differential equations that describe the behavior of a material under the influence of various physical fields. This method provides an accurate solution for simple geometries and idealized conditions.

The semi-analytical method combines analytical methods and numerical methods for solving differential equations. This method uses a wide range of numerical methods for solving equations for complex geometric shapes and inhomogeneous material properties. Using this method, it is possible to obtain a fairly accurate solution for different geometric configurations and material properties.

Finite element (FE) analysis is the most common method for modeling devices based on piezoactive materials. This method uses the decomposition of a complex geometric shape into simpler elements for which a solution can be obtained. The solution for each element is then combined to obtain a solution for the entire system. This method makes it possible to model the behavior of piezoactive devices with necessary accuracy, taking into account complex boundary conditions. However, it may require significant computational resources for complex full-scale models and non-stationary problems.

The main approach used in modeling bodies with multiphysical effects is the Hamilton principle, which includes the Gibbs free energy density. By taking into account the interaction of certain physical fields in this energy, it is possible to obtain a system of differential equations and boundary conditions that describe the behavior of piezoactive materials with various effects.

Further in this chapter, we consider recent works on each of the effects separately, as well as works that take into account several multiphysical effects simultaneously. Conclusions are drawn at the end of the chapter.

6.2. Piezoelectric Effect

The first effect to be considered is the piezoelectric effect. There are two types of piezoelectric effects: direct and inverse. The direct piezoelectric effect is manifested in the fact that when a piezoelectric material is mechanically deformed (for example, when it is compressed or stretched), a charge arises on its surface, which can be used to generate an electrical voltage. The inverse piezoelectric effect manifests itself in the fact that when an electric field is applied to a piezoelectric material, it deforms in the direction of the field. Thus, the inverse piezoelectric effect makes it possible to convert electrical voltage into mechanical deformation. The main advantages that have provided such a wide use of piezoelectric materials are good electromechanical properties, durability, relative ease of manufacture, as well as flexibility in the process of designing and integrating devices based on them. In this regard, the main areas of application are structural health monitoring, the emission and reception of acoustic waves, active suppression of parasitic vibrations, delay lines, piezoelectric motors and actuators, various sensors for measuring mechanical quantities, as well as energy harvesting. Energy harvesting means the conversion of free, unused energy of mechanical vibrations present in structures into electrical energy, and its subsequent accumulation. An overview of earlier works and various applications of piezoelectric transducers are given in Refs. [67, 70, 98, 114, 352, 358].

Reference [62] presents a new piezoelectric generator (PEG) based on the conventional impact frequency up-converted PEG, with a nonlinear magnetic force used to control the operating frequency. Distributed parameter modeling and simplified magnetic force modeling are used to analyze dynamic performance. Theoretical and experimental results show that the proposed PEG provides a lower initial frequency and a wider operating bandwidth. The maximum output power of the proposed PEG was 0.491 mW.

The authors of Ref. [71] presented a study of the optimal control of the active vibration of a piezoelectric cantilever beam. They used first-order shear deformation theory to obtain the flexural stiffness of the beam, created a finite element model based on

Hamilton variational principle and reduced its order using the modal superposition method. They placed sensors and actuators in pairs at the point of maximum deformation and used an independent modal space control method to actively control the beam structure. Numerical modeling and experiments have shown sufficient efficiency of the proposed model and control method.

Reference [451] proposes three different approaches in FE modeling of a piezoelectric bimorph cantilever, including one-dimensional (1D), two-dimensional and three-dimensional (3D) models. The accuracy of each method is discussed and compared with the experimental results. The study also discusses a comparison of the performance of various configurations of piezoelectric bimorph cantilevers, demonstrating that finite element modeling with COMSOL is simple and convenient for complex topological structures.

The authors of Ref. [203] propose a mathematical model based on the Euler–Bernoulli theory of continuous beams for a folded PEG with two degrees of freedom (DOF). The model was validated by finite element analysis and experimental measurements. The verified model is used for parametric analysis, from which useful observations are made for the design of high-performance PEGs.

Reference [235] presents a consistent geometrically nonlinear model (CGNM) of cantilever PEGs that takes into account high-order terms of the first derivative of deflection and the effect of geometrical nonlinearity on tip mass motion. The proposed CGNM is compared with the published geometrically nonlinear model (GNM) and experimental measurements, showing that the CGNM is more accurate, especially at a high level of excitation. Numerical simulations using CGNM also show that the optimal load resistance increases and the resonant frequency decreases with increasing the excitation level due to geometric nonlinearity.

Reference [240] presents a unified PEG model, based on equivalent circuit analysis, using an impedance electromechanical analogy. The model is used to analyze the behavior of three representative power harvesting interfaces: resistive, standard and synchronized switch harvesting on inductor (SSHI). The study also introduces the concept of a power limit and derives analytical expressions for the critical interface link, which is the minimum link required to reach the power limit and determines the link state of the system. The SSHI interface has been shown to have the lowest critical coupling and superior energy harvesting capability for weak coupling systems.

Reference [355] provides an improved dynamic model of piezo-electric stick-slip actuators, which are difficult to model due to their inconsistent stepping in forward and backward directions. The step inconsistency is due to the variable response of the piezoelectric power feeding caused by the behavior of the contact between the piezoelectric package and its preload mechanism. The proposed model includes a non-smooth contact model, which has been verified through simulations and experiments, during which a good fit of the stepping curves of the slider in bidirectional motion was observed. This model offers a new perspective for explaining and modeling the step inconsistency of stick-slip actuators and can help in their design and control for bidirectional motion.

The authors of Ref. [121] investigate the design of strongly coupled piezoelectric cantilevers to harvest vibration energy to power wireless sensors. An analytical approach based on the Rayleigh–Ritz method and a 2-DOF model is used to generate design recommendations that are experimentally validated. Three prototypes are presented, demonstrating some of the strongest quadratic global k^2 electrome-chanical coupling coefficients and wide bandwidth of PEG. One of the prototypes is capable of harvesting enough power (more than $100\,\mu\text{W}$) to power a wireless sensor node at up to 21% bandwidth.

Reference [198] explores the design and optimization of PEG under real-time stochastic oscillations using step-by-step guidelines from an electrical and mechanical perspective. The study highlights the importance of the contact layer and the correct visco-structural combined damping model for accurate power estimation. The presented method is applied to real-time stochastic oscillations, producing an electrical power of $1.32\,\text{mW}$ with a density of $495.92\,\mu\text{W/cm}^3$, demonstrating the potential of using PEG in autonomous sensors.

Reference [2] is devoted to the study of a mathematical model of a piezoelectric cantilever beam with a tip mass, subjected to vortex vibrations for sustainable energy harvesting. The study considers both unimorphic and bimorphic configurations and develops an equivalent mechanical model with one DOF integrated with an electrical model of a piezoelectric system. The article also developed a random process model for estimating the energy, obtained from stochastic excitation due to vortex formation. Moreover, expressions for the average power were obtained in an analytical form in this article. The results show that an optimal inductor constant exists for models with an inductor that meet the maximum average power condition.

Reference [181] presents an FE model using COMSOL software to simulate the dynamic mechanical and electrical behavior of a piezoelectric macrofiber composite (MFC) on carbon fiber composite structures for energy harvesting, sensing and actuation in various industries. The model integrates the physics of piezoelectric devices with electrical circuits and uses vibration data files for more accurate modeling. The simulation results were experimentally validated with a deviation of less than 10%, which demonstrated the applicability of the proposed FE model for the design and optimization of intelligent PEG composite structures.

Reference [280] examines the potential of piezoelectric crystals ($BaTiO_3$, $PbTiO_3$, $PbZrTiO_3$, PZT-5A and PZT-5H) in harvesting energy from vehicle suspension systems, while conducting comprehensive simulations to quantify the generated power. The authors developed a linear FE of a piezoelectric disk and integrated it into a full dynamic model of a car with 7-DOF. Also, the authors solved the equations of motion in the MATLAB environment. The results show that there is potential for significant energy harvesting from the wasted vibration of suspension systems, which can be used to power embedded electronic devices.

In Ref. [119], the authors focus on improving the performance of piezoelectric sensors for vibration analysis in industry. The principle of operation of piezoelectric accelerometers is implemented by researchers into a mathematical model, which is verified through experiments. The verified model is used to propose a new sensor design that aims to provide more accurate results and vibration information. The results of a comparative analysis with known data show that the choice of damping factor plays a decisive role in the development of accelerometer parameters and the improvement of the vibration analysis method.

Reference [160] is devoted to the use of piezoelectric bistable inertial generators to collect vibration energy. Their main purpose is to improve the power supply of isolated wireless sensors. The idea is to optimize the oscillator by examining the effect of various parameters such as load resistance, mass, stiffness and buckling level on its frequency response. The study develops a new analytical model of piezoelectric bistable oscillators based on a recent model of electromagnetic bistable oscillators. The model includes the study of subharmonic behavior and the stability robustness criterion to improve the prediction of experimental observations. The received optimization recommendations are confirmed by experimental data.

Reference [356] presents a transfer function, based on an analytical model for PEGs, used in wind power harvesting. The model is verified experimentally for two types of piezoelectric sensors, various wind conditions and geometric features. The model shows the potential for flutter energy harvesting and limit cycle oscillations, which are important in aerospace applications such as power harvesting from drones. The results show that the PZT-5A harvester generates more power during limit cycle oscillations compared to the PIC-255 generator. The proposed model is recommended for conventional wind energy harvesting and flutter harvesting.

Reference [207] proposes a PEG design that can convert environmental vibration into electrical energy for medical and industrial applications. In order to satisfy the broadband natural vibration frequency of the environment, it is proposed to control the tilt angle of the tip mass. The proposed design provides higher output power and wider operating bandwidth than conventional generator arrays. The measured output power is 2–3 mW in the frequency range from 19 Hz to 29 Hz and 9 mW in the frequency range from 290 Hz to 330 Hz. The proposed approach is modeled using FE analysis and analytical methods. The FE model is validated by experimental results.

Reference [155] aims to optimize the location of the masses and the piezoelectric patch for harvesting energy on a vibrating cantilever beam using a genetic algorithm. The amplitude of the electrical voltage, generated by the piezoelectric plates, is optimized by selecting the proof mass, moment of inertia, location, position of the piezoelectric plates, and point of application of the force as parameters. An analytical approach is proposed for estimating the amplitude of the electric voltage and a neural network is trained to obtain an approximate function of natural frequencies, based on the attachment parameters. The trained network is then used in a genetic algorithm to find the best optimization variables for any excitation frequency. Numerical simulation confirms the calculated electrical voltage using an analytical approach. The optimized design maximizes output energy by matching the natural frequency of the beam to the excitation frequency and outperforms the conventional PEG configuration at high frequency excitation.

In Ref. [418], a mathematical model of a vibrating plate with a piezoelectric patch, directly attached to it, is developed for energy harvesting in aviation applications, where cantilever PEGs cannot be installed. The model estimates the power, generated by the PEG under broadband excitation, and proposes the optimal position of the piezoelectric patch depending on the patch vibration modes. PEG performance is compared to a console generator set to the frequency of the most excited patch mode.

The authors of Ref. [127] present an application of the proper orthogonal decomposition method to model scenarios involving nonlinearities in microelectromechanical systems, including geometric and electrostatic nonlinearities. The method uses a low-dimensional subspace spanned by regular orthogonal modes to efficiently reduce polynomial terms to the cubic order, associated with large device displacements. Electrostatic nonlinearities are modeled using pre-computed manifolds based on the amplitudes of electrically active orthogonal modes. The method is being tested on various complex applications such as resonators, micromirrors and arches with internal resonances using classical time matching schemes and more advanced harmonic balance approaches. The study also discusses the robustness of the method in terms of estimating the frequency response function of selected output quantities of interest.

Reference [450] proposes an electromechanical step-reduction method for modeling piezoelectric composite beams of variable cross sections under conditions of longitudinal vibration, which takes into account the characteristics of electrical signal transmission and electromechanical coupling. The transfer equations relating the input and output state vectors are created, based on the proposed method, and confirmed on the example of a piezoelectric actuator. An actuator prototype was also made for experimental verification. The proposed method turned out to be more suitable for modeling piezoelectric composite beams of variable cross sections and optimizing the excitation of longitudinal vibrations in them.

The authors of Reference [375] present an applied theory of cylindrical bending vibrations of a bimorph plate, which takes into account the nonlinear distribution of the electric potential in piezoelectric layers. They developed a quadratic distribution of the electric potential and show that it allows one to accurately predict the resonant frequencies and vibration modes of a bimorph plate, as well as its forced vibrations under mechanical excitation. The theory was validated by comparing the results with FE analysis. The proposed method can be used to optimize the design of piezoelectric devices and systems.

In Ref. [365], the resonant frequencies of longitudinal vibrations of a complex system consisting of piezoceramic, elastic and acoustic elements are studied, with an emphasis on ultrasonic cutting devices. The study uses COMSOL and ACELAN FE software for comparison and evaluates the effect of geometry and dynamic viscosity on the first natural frequency. A modal and harmonic analysis of the system was carried out, the results of which show that high-frequency longitudinal oscillations of the rod element depend significantly on the dissipation coefficient of elastic elements and weakly on the viscosity of the contacting acoustic medium. The data obtained can be useful in the development of ultrasonic cutting medical devices.

The main results of Reference [373] include the development of applied theory and FE models for nonuniformly polarized piezoceramic materials in PEG applications. The methods used to obtain these results include the development of an applied second-order theory, based on the beam bending model, as well as the use of axisymmetric and planar FE models and numerical experiments in the ACELAN software.

Reference [290] is devoted to the analytical and numerical simulation of a nonuniformly polarized PEG, based on a porous piezoceramic plate. The effective properties of the piezoceramic material were calculated by the FE method in the ACELAN-COMPOS software. An applied theory for piecewise uniformly polarized PEG was developed and compared with the FE modeling results. The study proposes an efficient cantilever-based PEG design and analyzes the effect of percent porosity and boundary conditions on output performance. The paper also investigates various boundary conditions for specific piezoceramic materials with nonuniform polarization in the search for an efficient mounting scheme.

Reference [379] focuses on modeling shear vibrations of piezoelectric transducers with different porosities. The main task is to study the effect of porosity on the electromechanical coupling coefficients and the output electric potential. The effective properties of the porous ceramic materials were obtained using the ACELAN-COMPOS FE software, and the FE analysis of the transducer was performed using the COMSOL Multiphysics software. The study showed that the electromechanical coupling coefficient decreased with increasing porosity, but the value of the output electrical potential increased. This suggests that the use of porous piezoceramics may be beneficial for certain applications, requiring a high electrical output potential.

The main results of Ref. [85] are related to the numerical simulation of a piezoelectric composite consisting of porous piezoceramic rods regularly arranged in an elastic matrix. The study used a multilevel approach to calculate the effective moduli of porous piezoceramics and to analyze 1–3 piezocomposites with rods having calculated uniform properties. The effective properties of the 1–3 composites are determined for different percentages of porosity of piezoceramic rods, and the electromechanical properties are analyzed for different models of transducers made from the proposed composites. The methods used in this study included the homogenization method based on the Hill lemma and the FE method, as well as approximate analytical models.

Reference [382] presents the results of experimental and FE simulations of a cantilever-type PEG with a proof mass and an active base. The PEG is excited by a rotating actuator with magnetic devices. The generator itself consists of plate-type elements that experience bending deformations and cylindrical-type elements that experience compression deformations. The article describes the installation of the drive and shows the output characteristics of the generator at low frequency load.

The authors of Ref. [58] describe the use of the multi-objective Pareto optimization method to improve the efficiency of PEG. They solved an optimization problem for two types of generators: cantilever and stack. The cantilever generator optimization process was divided into several stages, which significantly reduced the amount of calculations. To solve both problems, the FE method was used, and it was found that this method is more suitable for optimizing the design of cantilever generators than stack generators, under given constraints.

Thus, the main methods for modeling oscillations of the devices, based on the piezoelectric effect, are semi-analytical models with lumped and distributed parameters, as well as equivalent models by using transfer functions and FE modeling. These methods have been successfully applied in the various studies cited above to model and optimize the operation of piezoelectric devices. The main directions of development in the field of piezoelectric transducers are the optimization of existing structures, the development of more complex structures and the search for new materials, including composite and functionally graded ones. Overall, advances in modeling techniques and materials science will continue to drive innovation and development of piezoelectric devices in the future.

6.3. Flexoelectric Effect

The second effect to be considered is the flexoelectric effect. This effect occurs when a piezoactive material is bent or deformed and consists in the appearance of an electric field. The flexoelectric effect is based on the distortion and change in the symmetry of the crystal lattice of the material due to the gradient of mechanical deformation, which leads to the appearance of internal charges and polarization. By this, it differs from the piezoelectric effect, which is based on a change in the distribution of charges in the crystal lattice of a material during mechanical deformation or the application of an external electric field. Moreover, the flexoelectric effect can manifest itself in various types of dielectrics, including not only crystalline but also amorphous, polycrystalline and even liquid materials. A similar effect, as a rule, manifests itself at the micro and nano scales in such bodies as thin films, membranes or nano dispersed systems. An overview of earlier works and various applications of flexoelectric transducers is given in Refs. [4, 48, 157, 405, 442, 505, 506].

Reference [443] investigates the electromechanical responses of nanoplates with piezoelectric and flexoelectric effects using Kirchhoff theory of thin plates. The study is devoted to the influence of the dynamic flexoelectric effect on the natural frequencies of a piezoelectric nanoplate. Exact frequency equations are obtained for various boundary conditions. The results show that the dynamic flexoelectric effect is size dependent and is more pronounced for higher vibration modes. Positive and negative flexoelectric coefficients affect natural frequencies in different ways, and the piezoelectric effect does not change the natural frequencies of free vibrations of a homogeneous nanoplate.

The authors of Ref. [420] investigate the nonlinear oscillations of a cantilever nanoactuator with a flexoelectric effect. The nonlinear control equation is derived by using nonclassical continuum mechanics and the Hamilton principle. The Galerkin method is used to reduce an equation to a system of ordinary differential equations. The resulting model is solved by the perturbation method for the response to free oscillations, and a state-space model for tracking control is derived using the sliding mode control algorithm and the Lyapunov stability theory. Numerical simulations show the effectiveness of the proposed control algorithm, which outperforms linear controllers such as fuzzy controllers.

Reference [333] investigates nonlinear random oscillations of a porous functionally graded nanobeam on a viscoelastic foundation by the method of statistical linearization. The beam is modeled using nonlocal Euler–Bernoulli beam theory with von Karman nonlinearity, and the Galerkin method is used to discretize the governing equation. The statistical linearization method has been found to give reliable results with excellent agreement with the analytical results. Numerical analysis shows that the RMS value of the fluctuations increases with increasing porosity, power-law index, nonlocal parameter, or mean value of the input data.

Reference [29] presents analytical solutions for a series-connected bimorph PEG with flexoelectric and nonlocal effects. The constitutive equations are derived using the extended Hamilton principle, and closed-form analytical solutions are obtained for non-periodic and harmonic base excitations. The study considers PEG with both the piezoelectric effect and the flexoelectric effect, and the results show that account of the flexoelectric effect significantly increases the output voltage, current, and power density, while reducing the speed of the end of the beam. The findings suggest that the flexoelectric effect is an important aspect for the efficient use of bimorph PEGs in advanced applications such as smart nanosensors.

Researchers in Ref. [494] presented a model of a porous axial functionally graded flexoelectric nanobeam, based on the Euler–Bernoulli theory, which accounts for the strain gradient. The model uses a modified power formula to take into account the volume fraction of porosity and considers two models of porosity distribution. The generalized differential-quadrature method is used to discretize the governing equations. Static bending deflections and frequencies of free oscillations are calculated for various boundary conditions. In the work, the authors confirm the mathematical formulation and numerical solution by comparing their results with the known ones and investigate the influence of various parameters on the behavior of a porous flexoelectric nanobeam.

Reference [27] considers nonlinear forced vibrations of viscoelastic flexoelectric nanobeams. The coupled governing equations of viscoelastic flexoelectric nanobeams are derived using Hamilton principle, non-classical continuum theory, and the Euler–Bernoulli beam model. The equations are solved numerically for distributed loading and clamped–clamped boundary conditions. The results show that the size effect and the effect of the viscoelastic medium can increase the vibration frequency of the nanobeams. Moreover, it was found that the natural frequency of nanobeams outside a viscoelastic medium strongly depends on the dimensional parameters, and an increase in the length and thickness of a nanobeam reduces the frequency. Finally, the results show that the amplitude of the nonlinear oscillations increases with the increase in the flexoelectric effect.

Reference [15] investigates nonlinear vibrations of functionally graded nanobeams with flexoelectric and surface effects on a nonlinear Pasternak foundation. A nonlocal, non-classical model of a nanobeam with a flexoelectric effect is used, and the effects of surface elasticity, dielectricity, piezoelectricity and bulk flexoelectricity are considered. The governing equations are derived using the Hamilton principle based on first-order shear deformation beam theory and elasticity theory with a nonlocal deformation gradient. The differential quadrature method is used to calculate the nonlinear natural frequency and mode shape. Numerical results show that considering flexoelectricity reduces the bending stiffness of flexoelectric functionally graded nanobeams. Separately, the authors studied the influence of various parameters, including flexoelectric coefficients, residual surface stresses, nonlocal parameters, length scale effects, cubic nonlinearity coefficients, power gradient index and geometric dimensions, on the nonlinear vibrational characteristics of nanobeams.

Reference [489] proposes a ring-type flexo-piezoelectric generator, which consists of an elastic ring with a flexo-piezoelectric patch, fixed on its surface. The researchers determined the output voltage and power under closed circuit conditions. A comparison of the characteristics of flexoelectric and piezoelectric power generators was carried out, as well as a parametric analysis of the output power depending on parameters such as patch size, load resistance and patch thickness. The results show that the flexoelectric ring power generator is more efficient than the similar piezoelectric one in the bending mode.

The authors of Ref. [103] explore the optimal combinations of several flexoelectric actuators on a rectangular plate by using a neural network model. The physical model uses an atomic force microscope sensor to generate an electric field gradient in a flexoelectric patch, resulting in a flexoelectric control force and moment. The voltage caused by flexoelectricity is mainly concentrated near the probe, and the size and shape of the flexoelectric region has a limited effect on operation. Therefore, only the positions of the actuator can be considered as input to the neural network model. The neural network model allows one to quickly predict the driving effect of a large set of actuators in different positions and accurately analyze the optimal position of the actuator. This study provides insights into the design of flexoelectric actuators to control various structures.

Reference [13] analyzes the free oscillations of a multi-layer microbeam under the action of an electric field using the Euler–Bernoulli beam theory and a modified deformation gradient theory. The beam has a porous functionally graded layer and two flexoelectric face layers that are exposed to an external electric field. A modified silica aerogel foundation model is used to account for the effect of the elastic foundation on the beam. The governing equations of motion are derived and solved by the Navier solution method. The paper investigates the influence of various parameters, such as the length-to-thickness ratio, porosity index, flexoelectric loads, the small-scale parameter and foundation parameters, on the dimensionless beam frequency.

Researchers in Ref. [20] present a reduced micromorphic model of composite metamaterials, which takes into account the flexoelectric effect. The paper formulates the phenomenon of polarization at the microlevel using the method of variation of the functional internal energy and the virtual work, performed by external fields. The resulting field equations, boundary conditions and constitutive relations are presented, and a 1D application with a physical explanation of the results is demonstrated.

Reference [421] analyzes nonlinear free oscillations of pre-actuated clamped-free isotropic Euler–Bernoulli piezoelectric nanobeams. The governing equations of motion are derived from the size-dependent theory of piezoelectricity, and a more accurate model is developed with a higher-order curvature–displacement relationship. In the work, a nonlinear equation of motion was obtained, within which the Lindstedt–Poincaré method was used to analyze the nonlinear free vibrations of a pre-actuated nanobeam. The influence of the applied voltage, the length scale parameter and the flexoelectric coefficient on the static deflection, nonlinear natural frequencies and effective nonlinearities was also investigated. In the results obtained, one can single out interesting phenomena that arise when different values of the length scale parameter and the flexoelectric coefficient are combined.

In Ref. [482], the authors present an analysis of nonlinear free vibrations and post-buckling of a nanobeam with a flexoelectric effect. The study is based on the Ehringen differential model and the Timoshenko beam theory. The von Karman strain–displacement relation, Gibbs electrical free energy and Hamilton principle are used to derive the equations of motion. The multiple scales method is used to obtain closed-form solutions for nonlinear constitutive equations. The study shows that flexoelectricity has a significant effect on the free vibrations of nanobeams and should be taken into account when designing nanoelectromechanical systems.

The researchers in Ref. [146] present a deep learning method that explores the effect of flexoelectricity in nanostructures. A deep neural network algorithm is used to map the relationship between input data and the response of a material. The neural network model is trained and tested using a database created by solving the governing flexoelectricity equations using the NURBS-based IGA formulation. The authors demonstrate the capabilities of the proposed method in terms of accuracy and computational efficiency in optimizing the topology of composite energy conversion systems.

Researchers in Ref. [99] consider buckling and oscillation of functionally graded flexoelectric nanobeams using a new theory developed exclusively for flexoelectric nanomaterials. The theory includes the development of electromechanical coupling equations, based on the von Karman deformation, forming enthalpy equation and Hamilton principle. The pre-buckling, buckling and vibrations of functionally graded nanobeams in freely supported and clamped-clamped boundary conditions were investigated, taking into account the Euler–Bernoulli beam model. The researchers showed that the flexoelectric effect has a significant impact on the buckling and oscillation characteristics of functionally graded nanobeams.

Reference [184] presents a higher-order size-dependent beam model for analyzing post-buckling and free vibrations of liquid-transporting nanotubes. This model takes into account nonlocal stresses, strain gradient effects, surface energy effects and slip flow effects. The governing post-buckling and oscillation equations are derived by using the Hamilton variational principle, and the two-step perturbation method is extended to the nonlinear analysis of fluid-transporting tubes. The study shows the influence of nano-effects on the nonlinear behavior and highlights the importance of comprehensively considering the effects of surface energy, small scale and slip flow in the nonlinear analysis of liquid-transporting nanotubes.

The authors of Ref. [271] investigate the effect of flexoelectricity on the wave dispersion characteristics of piezoelectric nanotubes using the Reddy higher-order shear deformation beam theory in combination with the nonlocal deformation gradient theory. The Hamilton principle is used to derive the governing equations and analytical solution of the eigenvalue problem. The work also studies the influence of various parameters, such as flexoelectric effects and applied voltage, on the dispersion curves of waves, as well as the dependence of the flexoelectric effect on dimensions.

Reference [284] considers the vibrational response of a homogeneous nanobeam with a functionally gradient base and a dielectric layer with both piezoelectric and flexoelectric properties, using local/nonlocal elasticity without paradoxes. To formulate the problem, the energy method and the Hamilton principle are used. Also, to solve complex equations, the generalized differential-quadrature method is applied. As part of the work, the influence of various parameters on the oscillatory behavior are investigated and it is shown that small-scale flexoelectricity is dominant in electromechanical coupling.

Reference [395] explores the influence of the effects of nonlocal elasticity theory on flexoelectric nanosensors, which have important applications in sensors, actuators and energy harvesting. In the study, the governing equations and boundary conditions are obtained using the generalized variational principle based on the electric Gibbs free energy for flexoelectric nanobeams under typical external loads. Expressions are obtained in analytical form for deflection and induced electric potential values. The numerical results show that nonlocal effects have a significant impact on the induced electric potential of flexoelectric sensors under the influence of transverse loads. The authors conclude that it is important to consider nonlocal effects for understanding and designing basic nanoelectromechanical components, subjected to various external loads.

Reference [486] presents a study of static bending and oscillations of a flexoelectric beam structure with a linear elastic substrate in a magnetic field. The basic equations and boundary conditions are derived using the Euler–Bernoulli beam theory and the Hamilton variational principle, based on the electrical Gibbs free energy density. Analytical expressions for the deflection and induced electric potential of the beam structure, as well as the natural frequency of the beam in open-circuit electrical conditions with surface electrodes, are obtained. The results show that the flexoelectric effect, a linear elastic substrate and a magnetic field have a significant effect on the behavior of a flexoelectric beam during static bending and oscillation, which can be useful in the design and development of flexoelectric devices with elastic substrates.

The authors of Ref. [220] study electromechanical coupling in flexoelectric circular plates, which is important for designing microelectromechanical systems and nanoelectromechanical systems such as sensors, actuators and energy harvesting devices. The analysis includes consideration of higher-order strain gradients and nonuniform electrical potential distribution. Analytical solutions show that the dynamic regimes of flexoelectric circular plates differ from those of piezoelectric circular plates due to inversion symmetry breaking, caused by the strain gradient. This study provides a fundamental understanding of the electromechanical coupling in flexoelectric circular plates and may be useful in the development of new flexoelectric devices such as flexoelectric mirrors.

The aim of Ref. [257] was to study the effect of flexoelectricity on a piezoelectric nanobeam, taking into account internal viscoelasticity, which was not previously considered. The frequency equation was derived using Lagrangian deformations and the Hamilton principle under closed-circuit conditions. An elastic nonlocal strain gradient model was used to calculate natural frequencies at the nanoscale, while the internal viscoelastic coupling was determined by a linear Kelvin–Voigt viscoelastic model. Analytical methods were used to determine the frequencies. The results of the work showed that the viscoelastic coupling can have a direct impact on the flexoelectric properties of the material.

In the framework of Ref. [32], the authors set themselves the task of studying free vibrations and static torsion of a flexoelectric micro/nanotube, taking into account the effect of independent polarization. Using the non-classical theory of continuum mechanics, based on the strain gradient, the governing equations for the torsion of flexoelectric micro/nanotubes are developed and a formulation with classical and non-classical boundary conditions is presented. In the article, polarization is introduced as a new variable in the equations, and its influence is investigated as an independent variable of the electric field. The results show that the parameters of the mechanical and electrical size effect have a great influence on the results, and polarization plays an important role in modeling torsional electromechanical structures.

Reference [415] highlights the potential for improving the performance of micro- and nanoscale power generators by exploiting design features such as curved shapes in addition to materials and topology. The authors developed a model of a piezo-flexoelectric curved beam, based on couple stress, to analyze the nonlinear frequency response of a nanosized arc-shaped beam. Their results show that the nonlinear vibration of energy harvesting devices is significantly affected by curved shapes, and that performance can be improved by tuning the curvature of the design. This work provides insights into the potential of curved designs to improve the efficiency and power density of energy harvesting devices in micro- and nanoelectromechanical devices.

The main result of Ref. [435] is the study of free vibrations of a nanoplate on a viscoelastic substrate, taking into account the effect of flexoelectricity. The equilibrium equation for the plate is derived based on the Hamilton principle. The natural frequency of the nanoplate is determined by using the analytical form of the solution. It has been established that the presence of the viscoelastic component of the elastic foundation is the main factor affecting the natural frequency of the nanostructure. The reliability of the study is confirmed by numerical calculations, taking into account factors such as temperature, size effect and compression force. This study provides important information about the behavior of nanoplates made of flexoelectric and viscoelastic materials under various loads, which is necessary for their practical use.

Reference [483] presents a nonlinear dynamic model of a flexo-electric cylindrical nanoshell, based on modified couple stress theory, which takes into account the size-dependent flexoelectric effect. In the work, the governing equations of nonlinear oscillations of the shell were derived. Also, the natural frequency and the generated electrical voltage were calculated. In addition, a study of the influence of geometric dimensions, character material length, and the amplitude of vibrations on the natural frequency of the nanoshell and the generated electrical voltage was made. The results show that the modified couple stress theory and the theory of large deformations are related to each other. Therefore, when analyzing natural frequencies and electromechanical behavior, it is necessary to take into account nonlinearity.

A relatively new direction in this area is the study of rotating nano and microbeams, which in the future can be used as gyroscopes and accelerometers.

Reference [156] investigates the simultaneous influence of flexo-electricity and piezoelectricity on the oscillatory behavior of a rotating piezoelectric microbeam using the Euler–Bernoulli beam theory and von Karman relations for deformation. The Gibbs energy function includes coupled stress, piezoelectric and flexoelectric effects.

The Hamilton principle is used to obtain the equations of motion and boundary conditions. The paper also explores how slenderness factors, changes in rotational speed and coefficients related to flexoelectric and piezoelectric properties affect the vibrations of a rotating microbeam. The data obtained indicate that the natural frequency increases with increasing rotation speed, and the use of piezoelectric and flexoelectric effects increases the rigidity of the structure, thereby increasing its natural frequency.

The main result of Ref. [330] is the study of bending vibrations of a flexoelectric microbeam rotating around its axis. It has also been proposed to use such a beam as a gyroscope to determine the angular velocity due to the electrical response, created by the Coriolis effect, associated with rotation. The authors derived a system of 1D equations and used theoretical analysis to demonstrate the possibility of using a flexoelectric beam as a gyroscope. The proposed flexoelectric beam gyroscope has a simpler design, compared to conventional piezoelectric beam gyroscopes.

In Ref. [416], the authors present a study of the response of static bending and free vibrations of rotating piezoelectric nanobeams, taking into account flexoelectric effects and geometric imperfections. In the study, the FE method was combined with the third-order beam shear deformation theory, and the structures were placed on Pasternak elastic foundations. The proposed approach shows high accuracy in comparison with other known works. The study demonstrates that the mechanical responses of rotating nanobeams are not similar to those under normal conditions, when the rotation speed is zero, which can be used as a benchmark for the practical design of nanoscale beam structures.

A review of works on modeling oscillations of flexoelectric devices showed that the main approaches used in the modeling process include the Hamilton principle, Euler–Bernoulli and Timoshenko beam theories, taking into account shear deformations, as well as various nonlinear effects. Further research in this area is aimed to developing new, more advanced models that can take into account more complex geometry and nonlinear properties of materials.

6.4. Pyroelectric Effect

The third considered effect is the pyroelectric effect. The pyroelectric effect is the appearance of an electric charge on the surface of the material when its temperature changes. Pyroelectric devices can be used to create heat sensors, infrared detectors, security systems and other applications related to the detection and measurement of heat and infrared radiation. Since we are primarily interested in mechanical vibrations, further works will be given where the pyroelectric effect is taken into account together with the piezo-electric effect or flexoelectric effect. An overview of earlier works and various applications of pyroelectric transducers is given in Refs. [45, 225, 226, 340].

In Ref. [342], the influence of thermal loads on the active control of structural vibrations using piezoelectric materials in the aviation and aerospace industries was studied. The authors propose an FE model of a composite beam with a fully covered piezoelectric sensor and actuator, based on the theory of high-order shear deformation, and consider electric potential fields and a linear temperature field. In the work, the Hamilton principle was used to derive governing electro-thermomechanical equations, and a controller with negative velocity feedback was implemented for the actuator. The effect of temperature in the range from $-70°C$ to $70°C$ on the active control of the vibration of a composite beam is analyzed.

Researchers in Ref. [115] study the oscillation dynamics of a self-sustaining pyroelectric capacitor by using thermally controlled bimetallic microcantilevers for energy harvesting microdevices. The proposed design has such features as a pyroelectric capacitor built in parallel with the thermal reservoirs, symmetrically supported by two bimetallic cantilever beams. As a result, physical contact is not required, which reduces the risk of mechanical failure. The paper presents a physically based reduced order model that takes into account heat transfer between and within various components, as well as various forces, including the force of gas damping. This model was used along with an optimization algorithm to create optimal designs for a temperature range from $26°C$ to $38°C$. This made it possible to reach the density range of a collected power of 0.4–$0.65\,\mathrm{mW/cm^2}$.

Reference [297] discusses the design and simulation of a hybrid piezo-pyroelectric energy device as a low-power source that can harvest energy from both vibration and heat sources. SolidWorks was used to design the device, and ANSYS FE software was used for the numerical calculation. The study showed that both the force and temperature of heat affect the power output and electrical voltage of the device. Moreover, the air heat temperature gives a higher output electrical voltage and power than the force generated by the air flow.

Reference [211] is also devoted to the study and development of hybrid energy harvesting devices that can simultaneously or separately capture ambient vibrational and thermal energies and convert them into electrical energy using the same collection device. The considered devices are based on lead-free lithium niobate piezo-electric materials, which makes them nontoxic, and are designed to power autonomous sensors in vehicles. The dissertation includes several designs of hybrid energy harvesters, their modeling, fabrication and experimental results with an emphasis on operating conditions for low-frequency low-amplitude oscillations and slow temperature fluctuations.

Reference [131] presents a micromechanical model of a medium made of a fiber-reinforced piezothermoelastic composite. This paper investigates thermoelastic damping and frequency shift in micro/nanoscale piezothermoelastic beams under various boundary conditions using analytical methods. The article concludes that piezothermoelastic beams have superior electromechanical–thermal properties and higher quality factor compared to monolithic beams for a high fiber volume fraction, which was confirmed by comparison with the literature. The data obtained indicate that an appropriate value of the fiber volume fraction can be chosen to design and optimize the frequency-sensitive microelectromechanical and nano-electromechanical piezothermoelastic beams.

The authors of Ref. [217] presented the development of a thermal transducer for small-scale energy harvesting using multiphysics memory alloys that transfer their heat to a pyroelectric element to convert thermal energy into electrical energy. The study highlights the benefits of using multiphysics memory alloys that combine shape memory characteristics with ferromagnetic properties, simplifying device design and implementation. The approach of thermally decoupling the pyroelectric element from the alloy allows for faster cooling, accounting for a higher oscillation frequency and provides 3–9 times higher power density compared to using electromagnetic conversion through a coil. The use of nonlinear electrical interfaces can also result in a theoretical gain from 8 to 25 times.

Reference [120] investigates the effect of flexoelectricity on the thermoelectromechanical behavior of a functionally graded electropiezoflexoelectric nanoplate by using a modified flexoelectric theory and the classical Kirchhoff theory. By using the variational method and the principle of minimum potential energy, the authors for the first time obtained coupled governing nonlinear differential equations of a nanoplate and related boundary conditions. The behavior of nanoplates under various mechanical, electrical and thermal loads with different boundary conditions is analyzed, as well as direct and inverse flexoelectric effects are investigated. The results show that the stiffness of the nanoplate increases in the presence of flexoelectricity, while the deflection and electric potential generated across the thickness of the nanoplate decrease. In addition, the induced polarization decreases as the nanoplate is subjected to a temperature ramp.

The researchers in Ref. [187] present a vibrating beam thermal energy harvesting device that simultaneously uses both pyroelectric and piezoelectric effects under impact of small temperature changes at low frequency. The study establishes theoretically and experimentally the relative contribution of these two mechanisms and introduces methods for optimizing the phase relationship between the contributed currents. The prototype device presented in the paper provides high performance under optimal conditions in a fixed configuration, demonstrating the potential of this approach for harvesting thermal energy.

In Ref. [252], the authors consider linear and nonlinear thermoelectromechanical free oscillations of a multi-layer piezoelectric beam with graphene plates in the middle layer. The piezoelectric layers act as sensors and are attached to the top and bottom surfaces of the base layer. The nonlinear differential equation governing the system is solved using direct technique of multiple scales, and an analytical relationship is derived to estimate amplitude-dependent nonlinear natural frequencies. The study shows that the reinforcement of the beam with graphene plates can lead to static instability, but pyroelectricity can delay this possibility and alleviate the nonlinearity of the reinforcement. The effect of temperature, pyroelectricity and the mass fraction of graphene plates decreases with a large oscillation amplitude.

In Ref. [407], the authors analyze thermal free vibrations of multi-layer piezoelectric beams reinforced with carbon nanotubes, taking into account the pyroelectric effect. Carbon nanotubes are randomly oriented, and their agglomeration is considered. The Chebyshev–Ritz method is used to determine natural frequencies and investigate the effects of pyroelectricity. The results obtained by the researchers show that neglecting the pyroelectric effect reduces the first natural frequency and that agglomeration eliminates the possibility of static instability in a nanocomposite beam exposed to external heat. The study highlights the potential of carbon nanotube-reinforced nanocomposites for use in thermal imaging and energy harvesting.

Reference [480] reports the observation of a stationary flexoelectric effect in the hot-pressed ferroelectric ceramic PZT-19, which leads to the creation of a new thermally stable polarization after firing metal electrodes with different depths of mechanical damage of the surfaces of the plate. The results show that plates, made of such ceramics, have an asymmetric dielectric hysteresis loop and retain the values of the pyroelectric and piezoelectric effects in the ferroelectric phase and the bolometric effect in the paraelectric phase after repeated heating cycles. The methods used in the study include the theoretical evaluation of the pyroelectric effect, the firing of metal electrodes to opposite surfaces of the ceramic plate, the replacement of electrodes and the evaluation of resulting polarization and related effects.

Reference [1] aims to improve the piezoelectric and pyroelectric properties of polyvinylidene fluoride (PVDF) nanofibers by electrospinning using certain nanofillers such as graphene oxide, graphene and halloysite nanotubes. Various concentrations of these nanofillers were mixed with a solution of PVDF and electrospun into nanofibers. The pyroelectric properties of the samples were evaluated by immersing the sealed samples in hot water and ice, and the piezoelectric properties were evaluated by bending tests. It was found that the addition of nanofillers improves the piezoelectric and pyroelectric properties of the samples by increasing the β-phase in the nanofibers. The highest efficiency in terms of piezoelectricity and pyroelectricity was achieved with PEG containing 1.6 wt.% graphene oxide. The ratio of the piezoelectric coefficients to the pyroelectric coefficients turned out to be constant ($\sim$1.5) and did not depend on the type and content of the nanofiller. The study also examined the effect of external force and vibration frequency on the output voltage and found that an increase in compression force and vibration frequency leads to an increase in output voltage. Finally, the fabricated nanogenerator was integrated into the insole and elbow to explore its ability to collect energy from body movement.

A review of works on modeling oscillations of devices with pyroelectric effect showed that the researchers use similar approaches as in the case of piezoelectric and flexoelectric effects, including the use of semi-analytical models with lumped and distributed parameters, as well as FE modeling. The presence of the pyroelectric effect expands the scope of energy harvesting devices and increases their efficiency. In general, the study of pyroelectric effects in such devices has the potential for further development and application in various fields of science and technology.

6.5. Piezomagnetic and Flexomagnetic Effects

The last two considered effects are the piezomagnetic and flexomagnetic effects. Piezomagnetic and flexomagnetic effects are phenomena associated with the interaction of a magnetic field with a strain field in solids. With the piezomagnetic effect, the deformation of the crystal lattice of the material causes a change in the magnetic properties, namely, the appearance of magnetic induction in the material. With the flexomagnetic effect, the change in magnetic properties is caused by a strain gradient in the material. This is close to the flexoelectric effect. An overview of earlier works and various applications of piezomagnetic and flexomagnetic transducers is given in Refs. [94, 95, 285].

Reference [487] presents an analytical solution for the problem of size-dependent static bending, free vibrations and buckling of a curved flexomagnetic nanobeam. The authors use the deformation gradient theory and the Timoshenko curved beam model and derive the governing equations and boundary conditions, based on the Hamilton variational principle. They use the Navier method to convert differential equations to algebraic equations and analyze the effects of opening angle, aspect ratio and scale parameter on bending strain, free vibrations and stability. The results are compared and confirmed by previous studies, showing good agreement.

In Ref. [102], a nanoporous piezoelectric and piezomagnetic beam is considered as an energy harvesting device. The beam is represented as a multi-layer composite beam with a piezoelectric and piezomagnetic core and surface interface layers. The governing equations and analytical solutions are obtained using the theory of the surface effect and Boit porous elasticity. Numerical calculations show that the resonant frequency, output power density and output voltage can be controlled by adjusting the surface material parameters, porosity properties, solid–liquid interaction effects, geometric dimensions and composition of piezoelectric and piezomagnetic materials. Overall, this approach promises to improve energy harvesting in nanoscale devices.

The authors of Ref. [484] present a new model of a transversely isotropic Timoshenko magnetoelectroelastic beam, which takes into account the effects of the microstructure and base of the beam. The model was developed by using a variational formulation, based on the Hamilton principle, an extended modified couple stress theory and a two-parameter Winkler–Pasternak elastic foundation model. The problems of static bending and free vibrations of a simply supported transversely isotropic magnetoelectroelastic beam under the action of a uniformly distributed load are analytically solved. Moreover, numerical results are presented, showing the influence of the microstructure, base and magnetoelectroelastic coupling on the beam response.

Reference [256] presents a model for studying the effect of flexomagnetism on small-sized actuators. The model includes linear Lagrangian deformations, the Euler–Bernoulli beam approach and the extended Hamilton principle. The model captures the desired influence through stress-driven nonlocal elasticity theory. Numerical results have been obtained analytically, showing that the diameter of nanotubes can significantly affect the characteristics of the flexomagnetic effect. The study provides insights into the behavior of flexomagnetic materials on a small scale and may inform the design of future nanoscale actuators.

In Ref. [258], the buckling of a magnetic composite nanoplate under the action of a planar 1D magnetic field is studied, taking into account the concept of flexomagnetism and the resulting bending forces and moments. The theory of nonlocal strain gradient is used along with the classical theory of plates to determine the equation describing the stability of a nanoplate. The axial magnetic force is also investigated, and the critical bending load is determined for various support conditions. The influence of the nonlocal parameter, the aspect ratio of the sheet and the 1D magnetic field on the critical load is discussed. It has been found that the flexomagnetic response is more noticeable if the nanoplate has a rectangular shape with an aspect ratio of less than one.

The main emphasis in Ref. [259] is made by the authors on the study of the flexomagnetic behavior of a vibrating multiphysical square beam in finite dimensions, taking into account the rotational inertia resulting from shear deformation, which is a new concept. The study uses a weak form of nonlocal deformation theory and the Galerkin method of weighted residuals to determine the solution. The main result is that the authors showed how the rotational inertia of the shear deformation and the flexomagnetic effect can influence each other. This opens new horizons for research into the flexomagnetic effect, as the rotational inertia of shear deformation can directly affect the flexomagnetic characteristics of small-sized actuators.

The main attention in Ref. [265] is paid to the study of the energy harvesting characteristics of a functionally graded magnetoelectroelastic cantilever beam during transverse vibrations. The coupled governing differential equations are derived using a lumped 1-DOF model, Gauss law, Newton law and Faraday law. The article evaluates the influence of key parameters such as gradient index, functionally graded pattern, number of turns and resistance on the output response of the system. The results show that the parameters of the material and system significantly affect the ability to collect energy from induced vibrations.

Reference [262] examines the mechanical strength and flexomagnetic properties of piezomagnetic nanosensors during post-buckling by using a nonlocal strain gradient elasticity approach. The sensor model is a Euler−Bernoulli beam with different boundary conditions. The study shows that the flexomagnetic effect is significant for less flexible boundary conditions and that post-buckling failure occurs earlier if the numerical amounts of the nonlocal parameter and the strain gradient are small and large, respectively. In general, the article gives an idea of the behavior of piezomagnetic sensors at the nanoscale.

Reference [488] investigates the flexomagnetic effect in a simply supported piezomagnetic nanoplate using the Kirchhoff plate model, strain gradient theory and Hamilton variational principle. In this work, the governing differential equations and the corresponding boundary conditions are derived, and an analytical solution is obtained for the size-dependent static bending, free vibrations and buckling of a flexomagnetic nanoplate using the Navier method. The study shows that the flexomagnetic coupling has a significant effect on the piezomagnetic response of thin nanoplates with small dimensions, which decreases with increasing geometric dimensions. The paper also discusses the effect of aspect ratio, flexomagnetic factor and scale parameter on bending strain, free vibrations and stability. Verification of the results shows good agreement with previous studies.

The authors of Ref. [366] investigate the influence of the flexomagnetic effect on the deflection and rotation of a cantilever beam using the Timoshenko beam model. The governing equation for magnetic induction includes strain gradients to account for the size effect phenomenon in a piezomagnetic body. The principle of virtual work is used to derive the governing equations, and an analytical approach is applied to study the effect of flexomagnetism on the Timoshenko piezomagnetic beam. The results show that deflection and rotation decrease with flexomagnetism, and this decrease becomes more significant as the beam thickness decreases. The developed general beam model recovers to the classical Timoshenko beam model with vanishing flexomagnecity.

Reference [261] studies the stability of a porous Euler–Bernoulli nanobeam, loaded in the axial direction, taking into account the flexomagnetic properties of the material. The study combines the stability equation with a nonlocal strain gradient elasticity model and uses a sinusoidal Navier transverse deflection to achieve the critical buckling load. It was found that the flexomagnetic effect is more pronounced at small scales, regardless of the strong magnetic field, and for large thicknesses, the difference between the responses of piezomagnetic and piezoflexomagnetic nanobeams is insignificant.

Reference [14] is based on the analysis of magnetoelectroelastic bending and buckling of three-layer nanoshells of double curvature, based on the nonlocal theory of elasticity. The theory of sinusoidal transverse shear with two variables and deformations in thickness is used to develop kinematic relationships, and the principle of virtual work is used to derive the governing equations of bending and buckling. The Navier approach is applied to simply supported boundary conditions. Also in the work, a parametric analysis is carried out to study the influence of a nonlocal parameter and applied electric and magnetic potentials on the bending and buckling of nanoshells.

In Ref. [199], a new model is presented for analyzing the vibrational response of a three-layer sandwich structure, consisting of a functionally graded composite core, reinforced with graphene nanoplates and piezoelectromagnetic face sheets. The extended rule of mixture and the Halpin–Tsai model are used to predict the mechanical properties of the nanocomposite core. The vibrational response of the model is evaluated by using the theory of sinusoidal shear deformation of the plate and the viscous-Pasternak base. The results show that the temperature difference affects the stiffness of the structure and reduces the frequency, and the use of a square basis for the foundation results in the lowest frequency and stiffness.

In Ref. [325], the non-stationary bending of an electromagnetoelastic rod, made of a homogeneous isotropic conductor, is considered taking into account the initial electromagnetic field, the Lorentz force, Maxwell equations and the generalized Ohm law. Due to the complexity of the general model for the Bernoulli–Euler rod, simplified equations are used, and the electromagnetic field is considered as quasi-stationary. Trigonometric series expansions and Laplace transformations are used to obtain solutions in integral form with kernels as influence functions. The images of nuclei in the space of Fourier transformations in the spatial coordinate are found, and examples of calculations for a concentrated load are given.

Reference [377] presents an applied theory of transverse oscillations of a bimorph cantilever, made of piezoactive materials, in an alternating magnetic field. The theory assumes a quadratic distribution of electric and magnetic potentials over the thickness of the cantilever and calculates the stress–strain state, the distribution of electric and magnetic fields and natural frequencies. The results are compared with an FE model built in COMSOL Multiphysics and show good agreement except for the beam clamp and free end. This theory can be used as a model for energy harvesting devices in an external alternating magnetic field.

The authors of Ref. [36] developed an applied theory of transverse steady oscillations of a two-layer plate consisting of a piezoelectric layer and a layer with piezomagnetic properties. Equations and boundary conditions are obtained by using the variational principle, based on hypotheses on the distribution of mechanical quantities, and electric and magnetic potentials over the thickness of the plate. The first natural frequencies were calculated and compared with the frequencies obtained using the ACELAN FE software. The results showed good agreement.

Reference [35] presents the application of the variational principle for deriving equations and boundary conditions for transverse steady oscillations of a bimorph from piezoelectric and piezomagnetic layers. Bimorph eigenfrequencies are calculated and compared with the finite element device model in ACELAN, showing good agreement between the two methods. The results of the study can be useful in the development and optimization of energy harvesting devices, based on piezoelectric and piezomagnetic materials.

In Ref. [329], researchers consider electromechanical fields in a thin composite beam, consisting of a flexoelectric semiconductor layer sandwiched between two piezomagnetic dielectric layers, on which a magnetic field is applied. The study uses the macroscopic theory of piezomagnets and flexoelectric semiconductors and derives a 1D model from 3D equations. The results show the bending deformation, caused by the magnetic field, and the redistribution or movement of charge carriers to the top and bottom of the beam through the combined piezomagnetic and flexoelectric couplings. The study introduces a coupling coefficient that characterizes the strength of the effect, which assumes a maximum at a certain ratio of the thicknesses of the piezomagnetic and semiconductor layers.

In conclusion, it can be noted that for modeling devices taking into account the piezomagnetic and flexomagnetic effects, similar methods are used as for flexoelectric devices. In particular, the Hamilton principle, the Euler–Bernoulli beam theory and the nonlocal strain gradient elasticity approach are applied. However, the main difference is the presence of a magnetic field and a magnetic moment, which are also taken into account in the models. Thus, the study of transducers, considering piezomagnetic and flexomagnetic effects, makes it possible to expand the applicability of harvesting devices and increases their efficiency.

6.6. Mixed Multiphysics Effects

Finally, we look at the studies that consider problems in which all four types of fields interact: mechanical, electrical, temperature and magnetic. This is quite a difficult task, since there are no ready-made

FE packages that describe such a multiphysics interaction of effects yet. This leads to the need to solve rather cumbersome problems by semi-analytical methods.

Reference [455] presents a comprehensive analytical model and theoretical solution for a flexoelectric energy nanogenerator, driven by a time-harmonic magnetic field and a thermal field. The model includes coupled extensional and bending oscillations, flexoelectricity, surface effect and nonlinear characteristics of the magneto-mechanical-thermo-electric coupling. The key performance indicators of the generator can be monitored and adjusted by using external magnetic fields, prestressing and thermal field effects. The theoretical model and solution are suitable for both macroscopic and microscopic, linear and nonlinear mechanical analysis and provide the basis for research related to high-frequency vibration with shear deformation in devices.

The authors of Ref. [161] analyze free oscillations and buckling of magnetoelectroelastic functionally graded microplates in a thermal medium. Microplates are composed of two phases of piezoelectric and piezomagnetic materials distributed through the thickness based on a power law model. The study uses a modified strain gradient theory combined with a higher-order generalized shear strain theory and isogeometric analysis to derive equilibrium equations. The article presents benchmark solutions for oscillations and buckling of magnetoelectroelastic functionally graded microplates, as well as data on the influence of various parameters on dimensionless frequencies and critical buckling loads.

Reference [260] considers non-stationary thermoelastic coupling with free oscillations of piezomagnetic microbeams, taking into account the flexomagnetic effect. The generalized Lord–Schulman theory of thermoelasticity is used to analyze the interaction between elastic deformation and thermal conductivity. The paper concludes that the flexomagnetic effect is significant when the value of the thermal conductivity of a material increases or the thermal relaxation time of the heat source is shorter. This study could become the basis for further studies of magnetothermoelastic small-scale piezoflexo-magnetic structures, based on thermal conductivity models.

Reference [277] presents an analytical approach to the analysis of the thermal stability of piezomagnetic nanosensors and nanoactuators, taking into account flexomagnetic effects and geometric imperfections. The governing equations for strips of piezoflexomagnetic nanoplates, subjected to an external temperature load, are obtained by using the theory of shear deformation of the plate of the first order and the theory of nonlocal deformation gradient. The results obtained are supported by the existing literature and demonstrate that the calculated buckling and post-buckling temperatures are affected by various factors such as plate slenderness factor, initial midplane rise, temperature distribution and magnetic potential. The proposed analytical solutions and numerical results can serve as a guide for future analyzes of piezoflexomagnetic nanosensors and nanoactuators.

Owing to the bulkiness of such multiphysical problems, now there are few works devoted to their solution. However, taking into account more effects in the future may allow the development of a universal energy harvesting device.

6.7. Conclusions

This final chapter of the book, in particular, develops and extends the results presented in Chapters 4 and 5. It has reviewed some of the work in the last 5 years in the field of simulation of the vibration of devices based on piezoactive materials, which included five effects: piezoelectric, flexoelectric, pyroelectric, piezomagnetic and flexomagnetic. Analytical, semi-analytical and FE analysis turned out to be the main methods for modeling vibrations of devices operating on these effects. Various approaches are used, such as the Hamilton principle, Euler–Bernoulli or Timoshenko beam theories, shear strain theories of various orders, nonlocal strain gradient elasticity and couple stress theory. In the optimization problems of such devices, in addition to the classical parametric analysis and genetic algorithms, the active use of neural networks begins. This allows one to significantly reduce the time and cost of optimizing devices, because the neural network can independently determine the most optimal device parameters based on the specified performance criteria. In the future, this will make it possible to design universal piezoelectric transducers, both for the macro level and for micro- and nanoscales. Separately, we can note a relatively new direction in this area, associated with the study of rotating nano- and microbeams, which in the future can be used as gyroscopes and accelerometers.

Owing to their high efficiency and versatility, piezoelectric transducers can be used in a wide range of applications, from energy harvesting systems and medical technology to industrial robots and aerospace vehicles. Thus, we can conclude that this area will be actively developed in the near future and has great potential for creating new, more efficient and versatile transducers, based on piezoactive materials.

Bibliography

1. Abbasipour, M., Khajavi, R., Yousefi, A. A., Yazdanshenas, M. E., Razaghian, F. and Akbarzadeh, A. (2019). Improving piezoelectric and pyroelectric properties of electrospun PVDF nanofibers using nanofillers for energy harvesting application. *Polymers for Advanced Technologies*, 30(2), 279–291.
2. Adhikari, S., Rastogi, A. and Bhattacharya, B. (2020). Piezoelectric vortex induced vibration energy harvesting in a random flow field. *Smart Materials and Structures*, 29(3), 035034.
3. Agrawal, O. (2004). Application of fractional derivatives in thermal analysis of disk brake. *Nonlinear Dynamics*, 38, 191–206.
4. Ahmadpoor, F. and Sharma, P. (2015). Flexoelectricity in two-dimensional crystalline and biological membranes. *Nanoscale*, 7(40), 16555–16570.
5. Aifantis, E. C. (1994). Gradient effects at macro micro and nano scales. *Journal of the Mechanical Behavior of Biomedical Materials*, 5, 355–375.
6. Akopyan, V. A., Parinov, I. A., Zakharov, Y. N., Chebanenko, V. A. and Rozhkov, E. V. (2015). Advanced investigations of energy efficiency of piezoelectric generators. In *Advanced Materials — Studies and Applications*, eds. Parinov, I. A., Chang, S. H. and Theerakulpisut, S. (Nova Science Publishers, New York) pp. 417–436.
7. Akopyan, V. A., Soloviev, A. N., Parinov, I. A. and Shevtsov, S. N. (2010). *Definition of Constants for Piezoceramic Materials* (Nova Science Publishers, New York).
8. Akopyan, V. A., Zakharov, Y. N., Parinov, I. A., Rozhkov, E. V., Shevtsov, S. N. and Chebanenko, V. A. (2013). Optimization of output characteristics of the bimorph power harvesters. In *Nano- and Piezoelectric Technologies, Materials and Devices*, ed. Parinov, I. A. (Nova Science Publishers, New York) pp. 111–131.

9. Akopyan, V. A., Zakharov, Yu. N., Parinov, I. A., Rozhkov, E. V., Shevtsov, S. N., Wu, P. C. and Wu, J. K. (2013). Theoretical and experimental investigations of piezoelectric generators of various types. In *Physics and Mechanics of New Materials and Their Applications*, ed. Parinov, I., Chang, S. H. (Nova Science Publishers, New York) pp. 309–334.

10. Anderson, P. V. (1960). Qualitative considerations regarding the statistics of the phase transition in ferroelectrics. In *Physics of Dielectrics*, ed. Anderson, P. V. (Publishing House of the Academy of Sciences of the USSR, Moscow), p. 290 (in Russian).

11. Anton, S. R. (2011). Multifunctional piezoelectric energy harvesting concepts. PhD Thesis. Virginia Polytechnic Institute and State University, Blackburn, Virginia, USA, p. 190.

12. Anton, S. R. and Sodano, H. A. (2007). A review of power harvesting using piezoelectric materials (2003–2006). *Smart Materials and Structures*, 16, R1.

13. Arani, A. G., Zarei, H. B. A. and Pourmousa, P. (2019). Free vibration response of FG porous sandwich micro-beam with flexoelectric face-sheets resting on modified silica aerogel foundation. *International Journal of Applied Mechanics*, 11(9), 1950087.

14. Arefi, M. and Amabili, M. (2021). A comprehensive electro-magneto-elastic buckling and bending analyses of three-layered doubly curved nanoshell, based on nonlocal three-dimensional theory. *Composite Structures*, 257, 113100.

15. Arefi, M., Pourjamshidian, M., Ghorbanpour Arani, A. and Rabczuk, T. (2019). Influence of flexoelectric, small-scale, surface and residual stress on the nonlinear vibration of sigmoid, exponential and power-law FG Timoshenko nano-beams. *Journal of Low Frequency Noise, Vibration and Active Control*, 38(1), 122–142.

16. Arlt, G. and Neumann, H. (1988). Internal bias in ferroelectric ceramics: Origin and time dependence. *Ferroelectrics*, 87, 109–120.

17. Arrieta, A. F., Hagedorn, P., Erturk, A. and Inman, D. J. (2010). A piezoelectric bistable plate for nonlinear broadband energy harvesting. *Applied Physics Letters*, 97, 104102.

18. Ashraf, M. W., Tayyaba, S. and Afzulpurkar, N. (2011). Micro Electromechanical Systems (MEMS) based microfluidic devices for biomedical applications. *International Journal of Molecular Sciences*, 12, 3648–3704.

19. Ashraf, M. W., Tayyaba, S., Nisar, A., Afzulpurkar, N., Bodhale, D., Lomas, T., Poyai, A. and Tuantranont, A. (2010). Design, fabrication and analysis of silicon hollow microneedles for transdermal drug delivery system for treatment of hemodynamic dysfunctions. *Cardiovascular Engineering*, 10, 91–108.

20. Atef, H. M. and El-Dhaba, A. R. (2022). Modeling the flexoelectric effect via the reduced micromorphic model. *Composite Structures*, 290, 115504.
21. Athanassoulis, G. A., Belibassakis, K. A., Mitsoudis, D. A., Kampanis, N. A. and Dougalis, V. A. (2008). Coupled-mode and finite-element solutions of underwater sound propagation problems in stratified acoustic environments. *Journal of Comp. Acoustics*, 16(1), 83–116.
22. Aurivillius, B. (1949). Mixed bismuth oxides with layer lattices: I. Structure type of $CaBi_2B_2O_9$. *Arkiv för kemi*, 1(54), 463–480.
23. Aurivillius, B. (1949). Mixed bismuth oxides with layer lattices: II. Structure type of $Bi_4Ti_3O_{12}$. *Arkiv för kemi*, 1(58), 499–512.
24. Aurivillius, B. (1950). Mixed bismuth oxides with layer lattices: III. Structure type of $BaBi_4Ti_4O_{15}$. *Arkiv för kemi*, 2(37), 512–527.
25. Aurivillius, B. (1962). Ferroelectricity in the compound $BaBi_4Ti_5O_{18}$. *Physical Review Journals*, 126, 893–896.
26. Avazzadeh, Z., Heydari, M. and Loghmani, G. B. (2011). Numerical solution of Fredholm integral equations of the second kind by using integral mean value theorem. *Applied Mathematical Modelling*, 35(5), 2374–2383.
27. Bagheri, R. and Tadi Beni, Y. (2021). On the size-dependent nonlinear dynamics of viscoelastic/flexoelectric nanobeams. *Journal of Vibration and Control*, 27(17–18), 2018–2033.
28. Barton, D. A. W., Burrow, S. G. and Clare L. R. (2010). Energy harvesting from vibrations with a nonlinear oscillator. *Journal of Vibration and Acoustics*, 132, 021009.
29. Basutkar, R. (2019). Analytical modelling of a nanoscale series-connected bimorph piezoelectric energy harvester incorporating the flexoelectric effect. *International Journal of Engineering Science*, 139, 42–61.
30. Belokon', A. V. and Nasedkin, A. V. (1998). Modeling of ultrasonic wave piezo-emitters using the ANSYS software package. *News of Taganrog Radio-Technical University*, 10(4), 147–150.
31. Ben Jennet, D., Marchet, P., El Maaoui, M. and Mercurio, J. P. (2005). From ferroelectric to relaxor behaviour in the Aurivillius-type $Bi_{4-x}Ba_xTi_{3-x}Nb_xO_{12}$ ($0 \leq x \leq 1.4$) solid solutions. *Materials Letters*, 59, 376–382.
32. Beni, Y. T. (2022). Size dependent torsional electro-mechanical analysis of flexoelectric micro/nanotubes. *European Journal of Mechanics-A/Solids*, 95, 104648.
33. Berdonosov, P. S., Charkin, D. O., Dolgikh, V. A., Stefanovich, S. Yu., Smith, R. I. and Lightfootd, Ph. (2004). $Bi_{2-x}Ln_xWO_6$: A novel layered structure type related to the Aurivillius phases. *Journal of Solid State Chemistry*, 177, 2632–2634.

34. Betts, D. N., Kim, H. A., Bowen, C. R. and Inman, D. J. (2012). Optimal configurations of bistable piezo-composites for energy harvesting. *Applied Physics Letters*, 100, 114104.
35. Binh, D. T., Chebanenko, V. A., Duong, L. V., Kirillova, E., Thang, P. M. and Soloviev, A. N. (2020). Applied theory of bending vibration of the piezoelectric and piezomagnetic bimorph. *Journal of Advanced Dielectrics*, 10(03), 2050007.
36. Binh, D. T., Soloviev, A. N., Chebanenko, V. A., Kirillova, E. and Ha, T. H. D. (2021). Applied theory of bending vibration of magnetoelectroelastic Bimorph. In *Proceedings of the 2nd Annual International Conference on Material, Machines and Methods for Sustainable Development (MMMS2020).* (Springer International Publishing, USA) pp. 337–342.
37. Blake, S. M., Falconer, M. J., McCreedy, M. and Lightfoot, P. (1997). Cation disorder in the ferroelectric Aurivillius phases of the type $Bi_2ANb_2O_9$ (A = Ba, Sr, Ca). *Journal of Materials Chemistry*, 7, 1609–1613.
38. Blarigan, L. V., Danzl, P. and Moehlis, J. (2012). A broadband vibrational energy harvester. *Applied Physics Letters*, 100, 253904.
39. Blinc, R. and Žekš, B. (1974). *Ferroelectrics and Antiferroelectrics* (North-Holland Publishing Company, Amsterdam).
40. Boonma, A., Narayan, R. J. and Lee, Y.-S. (2013). Analytical modeling and ev aluation of microneedles apparatus with deformable soft tissues for biomedical applications. *Computer-Aided Design and Applications*, 10, 139–157.
41. Borg, S. and Svensson, G. (2001). Crystal chemistry of $Bi_{2.5}Me_{0.5}Nb_2O_9$ (Me = Na, K): A powder neutron diffraction study. *Journal of Solid State Chemistry*, 157, 160–165.
42. Borg, S., Svensson, G. and Bovin, J.-O. (2002). Structure study of $Bi_{2.5}Na_{0.5}Ta_2O_9$ and $Bi_{2.5}Na_{m-1.5}Nb_mO_{3m+3}$ (m = 2–4) by neutron powder diffraction and electron microscopy. *Solid State Chemistry*, 167, 86–96.
43. Borodin, V. Z., Bondarenko, E. I. and Sitalo, E. I. (1991). Formation of the oscillation spectrum of plates. In *Proceedings of XI All-Union Acoustic Conference.* (Moscow State University Press, Moscow), pp. 94–97 (in Russian).
44. Borodin, V. Z., Bondarenko, E. I., Sitalo, E. I. and Cisse, S. D. (1977). Spatially inhomogeneous piezostructures in ferroelectric ceramics. In *Piezoelectric Materials and Transducers.* (Rostov State University Press, Rostov-on-Don) pp. 167–172 (in Russian).
45. Bowen, C. R., Taylor, J., LeBoulbar, E., Zabek, D., Chauhan, A. and Vaish, R. (2014). Pyroelectric materials and devices for energy

harvesting applications. *Energy & Environmental Science*, 7(12), 3836–3856.

46. Bowen, C. R., Topolov, V. Y. and Kim, H. A. (2016). *Modern Piezoelectric Energy-Harvesting Materials*, Vol. 238. (Springer, Cham, Switzerland).

47. Bryden, I., Grinsted, T. and Melville, G. (2004). Assessing the potential of a simple channel to deliver useful energy. *Applied Ocean Research*, 26, 198–204.

48. Buka, A. and Éber, N. (2013). *Flexoelectricity in Liquid Crystals: Theory, Experiments and Applications* (World Scientific, Singapore).

49. Butkovskii, A. G., Postnov, S. S. and Postnova, E. A. (2013). Fractional integro-differential calculus and its control-theoretical applications. I. Mathematical fundamentals and the problem of interpretation. *Automation and Remote Control*, 74, 543–574.

50. Butkovskii, A. G., Postnov, S. S. and Postnova, E. A. (2013). Fractional integro-differential calculus and its control-theoretical applications. II. Fractional dynamic systems: Modeling and hardware implementation. *Automation and Remote Control*, 74, 725–749.

51. Cao, J., Wang, W., Zhou, S., Inman, D. and Lin, J. (2015). Nonlinear time-varying potential bistable energy harvesting from human motion. *Applied Physics Letters*, 107, 143904.

52. Cao, J., Zhou, S., Inman, D. J. and Chen, Y. (2015). Chaos in the fractionally damped broadband piezoelectric energy generator. *Nonlinear Dynamics*, 80(4), 1705–1719.

53. Carl, K. and Hardtl, K. H. (1978). Electrical after-effects in Pb(Ti, Zr)O, ceramics. *Ferroelectrics*, 17, 473–486.

54. Carpinteri, A., Cornetti, P., Sapora, A., Di Paola, M. and Zingales, M. (2009). Fractional calculus in solid mechanics: Local versus non-local approach. *Physica Scripta*, 2009, 014003.

55. Champarnaud-Mesjard, J.-C., Frit, B. and Watanabe, A. (1999). Crystal structure of $Bi_2W_2O_9$, the $n = 2$ member of the homologous series $(Bi_2O_2)B_n^{VI}O_{3n+1}$ of cation-deficient Aurivillius phases. *Journal of Materials Chemistry*, 9, 1319–1322.

56. Chang, S.-H., Parinov, I. A. and Topolov, V. Y. (eds.). (2014). *Advanced Materials — Physics, Mechanics and Applications*, Vol. 152 (Springer, Cham, Heidelberg, New York, Dordrecht, London).

57. Chebanenko, V. A., Akopyan, V. A. and Parinov, I. A. (2015). Piezoelectric generators and energy harvesters: Modern state of the art. In *Piezoelectrics and Nanomaterials: Fundamentals, Developments and Applications*. ed. Parinov, I. A. (Nova Science Publishers, New York) pp. 243–277.

58. Chebanenko, V. A., Zhilyaev, I. V., Soloviev, A. N., Cherpakov, A. V. and Parinov, I. A. (2020). Numerical optimization of the piezoelectric generators. *Journal of Advanced Dielectrics*, 10, 2060016.
59. Chen, G., Bai, W., Sun, L., Wu, J. and Ren, Q. (2013). Processing optimization and sintering time dependent magnetic and optical behaviors of Aurivillius $Bi_5Ti_3FeO_{15}$ ceramics. *Journal of Applied Physics*, 113, 034901.
60. Chen, L.-Q., Jiang, W.-A., Panyam, M. and Daqaq, M. F. (2016). A broadband internally resonant vibratory energy harvester. *Journal of Vibration and Acoustics*, 138, 061007.
61. Chen, S., Liu, H., Ji, J., Kan, J., Jiang, Y. and Zhang, Z. (2020). An indirect drug delivery device driven by piezoelectric pump. *Smart Materials and Structures*, 29, 75030.
62. Cheng, Q., Lv, Z., Liu, Z. and Wang, Q. (2022). Theoretical modelling and experimental investigation on a frequency up-converted nonlinear piezoelectric energy harvester. *Sensors and Actuators A: Physical*, 347, 113979.
63. Cherpakov, A. V. and Kokareva, Y. A. (2019). Modal analysis of the cantilever type piezo-electric generator characteristics with active based on numerical simulation. *IOP Conference Series: Materials Science and Engineering*, 698, 066020.
64. Cherpakov, A. V., Parinov, I. A. and Haldkar, R. K. (2022). Parametric and experimental modeling of axial-type piezoelectric energy generator with active base. *Applied Science*, 12, 1700–1721.
65. Cherpakov, A. V., Parinov, I. A., Soloviev, A. N. and Rozhkov, E. V. (2019). Experimental studies of cantilever type PEG with proof mass and active clamping. In *Advanced Materials — Proceedings of the International Conference on "Physics and Mechanics of New Materials and Their Applications", PHENMA 2018*, Vol. 224, eds. Parinov, I., Chang, S. H. and Kim, Y. H. (Springer, Cham, Switzerland) pp. 593–601.
66. Chilabi, H. J., Salleh, H., Al-Ashtari, W., Supeni, E. E., Abdullah, L. C., As'arry, A. B., Rezali, K. A. M. and Azwan, M. K. (2021). Rotational piezoelectric energy harvesting: A comprehensive review on excitation elements, designs, and performances. *Energies*, 14, 3098.
67. Chopra, I. (2002). Review of state of art of smart structures and integrated systems. *AIAA Journal*, 40(11), 2145–2187.
68. Choy, J.-H., Kim, J.-Y. and Chung, I. (2001). Neutron diffraction and X-ray absorption spectroscopic analyses for lithiated Aurivillius-type layered perovskite oxide, $Li_2Bi_4Ti_3O_{12}$. *The Journal of Physical Chemistry B*, 105, 7908–7912.

69. Chu, M. V., Cadles, M. T., Piffard, Y., Marie, A. M., Gauter, E., Joubert, O., Ganne, M. and Brohan, L. (2003). Evidence for a monoclinic distortion in the ferroelectric Aurivillius phase $Bi_3LaTi_3O_{12}$. *Journal of Solid State Chemistry*, 172, 389–395.
70. Covaci, C. and Gontean, A. (2020) Piezoelectric energy harvesting solutions: A review. *Sensors*, 20(12), 3512.
71. Cui, M., Liu, H., Jiang, H., Zheng, Y., Wang, X. and Liu, W. (2022). Active vibration optimal control of piezoelectric cantilever beam with uncertainties. *Measurement and Control*, 55(5–6), 359–369.
72. Cveticanin, L. (2009). Oscillator with fraction order restoring force. *Journal of Sound and Vibration*, 320, 1064–1077.
73. Cveticanin, L. and Zukovic, M. (2009). Melnikov's criteria and chaos in systems with fractional order deflection. *Journal of Sound and Vibration*, 326, 768–779.
74. Daqaq, M. (2011). Transduction of a bistable inductive generator driven by white and exponentially correlated gaussian noise. *Journal of Sound and Vibration*, 330, 2554–2564.
75. Dasgupta, S. S., Rajamohan, V. and Jha, A. K. (2018). Dynamic characterization of a bistable energy harvester under gaussian white noise for larger time constant. *Arabian Journal for Science and Engineering*, 44, 721–730.
76. De Araujo, C. A., Cuchlaro, J. D., Mcmillan, L. D., Scott, M. C. and Scott, J. F. (1995). Fatigue-free ferroelectric capacitors with platinum electrodes. *Nature*, 374, 627–629.
77. Debnath, L. (2003). Recent applications of fractional calculus to science and engineering. *International Journal of Mathematics and Mathematical Sciences*, 54, 3413–3442.
78. Dereshgi, H. A., Dal, H. and Yıldız, M. Z. (2021). Piezoelectric micropumps: State of the art review. *Microsystem Technologies*, 27, 4127–4155.
79. Desu, S. B., Cho, H. S. and Joshi, P. C. (1997). Highly oriented ferroelectric $CaBi_2Nb_2O_9$ thin films deposited on Si(100) by pulsed laser deposition. *Applied Physics Letters*, 70, 1393–1395.
80. Devonshire, A. F. (1949). Theory of barium titanate: Part I. *Philosophical Magazine*, 42, 1040–1063.
81. Devonshire, A. E. (1951). Theory of barium titanate: Part II. *Philosophical Magazine*, 42, 1065–1080.
82. Devonshire, A. F. (1954). Theory of ferroelectrics. *Advances in Physics*, 3, 85–130.
83. Di Paola, M and Zingales, M. (2008). Long-range cohesive interactions of non-local continuum faced by fractional calculus. *International Journal of Solids and Structures*, 21, 5642.

84. Dneprovski, V. G., Karapetyan, G. Y. and Parinov I. A. (2016). *Surface Acoustic Wave Devices* (Nova Science Publishers, New York).
85. Do, T. B., Nasedkin, A., Oganesyan, P. and Soloviev, A. (2023). Multilevel modeling of 1–3 piezoelectric energy harvester based on porous piezoceramics. *Journal of Applied and Computational Mechanics*, 9(3), 763–774.
86. Dong, X. W., Wang, K. F., Wan, J. G., Zhu, J. S. and Liu, J.-M. (2008). Magnetocapacitance of polycrystalline $Bi_5Ti_3FeO_{15}$ prepared by sol-gel method. *Journal of Applied Physics*, 103, 094101.
87. Drobotov, Y. E. and Vakulov, B. G. (2022). Smoothness properties of a Riesz potential type operator with logarithmic characteristic. *News of Universities of the North-Caucasus Region. Natural Sciences*, 1, 4–11.
88. Drobotov, Y. E. and Vakulov, B. G. (2023). On solvability of integral equations of the first kind with mild singularity in the kernel. In *Physics and Mechanics of New Materials and Their Applications — Proceedings of the International Conference PHENMA 2021–2022, Springer Proceedings in Materials*, eds. Parinov, I. A., Chang, S.-H. and Soloviev, A. N. (Springer Nature, Cham, Switzerland), Vol. 20, pp. 120–132.
89. Duong, L. V., Pham, M. T., Chebanenko, V. A., Solovyev, A. N. and Nguyen, C. V. (2017). Finite element modeling and experimental studies of stack-type piezoelectric energy harvester. *International Journal of Applied Mechanics*, 9, 1–16.
90. Duran-Martin, P., Castro, A., Millan, P. and Jimenez, B. (1998). Influence of Bi-site substitution on the ferroelectricity of the Aurivillius compound $Bi_2SrNb_2O_9$. *Journal of Materials Research*, 13, 2565–2571.
91. Dzhafarov, A. S. (1966). The best approximation to finite spherical sums and some differential properties of harmonic functions in the ball. In *Embedding Theories and Their Applications. Proceedings of the Symposium on Embedding Theory*, Baku, 1966. ed. Kudryavcev, L. (Nauka, Moscow) pp. 75–81 (in Russian).
92. Dzhafarov, A. S. (1985). Constructive description of generalized Besov classes on a multidimensional sphere. *DAN SSSR*, 285(3), 542–546 (in Russian).
93. Eichhorn, C., Tchagsim, R., Wilhelm, N. and Woias, P. (2011). A smart and self-sufficient frequency tunable vibration energy harvester. *Journal of Micromechanics and Microengineering*, 21, 1–11.
94. Eliseev, E. A., Morozovska, A. N., Glinchuk, M. D. and Blinc, R. (2009). Spontaneous flexoelectric/flexomagnetic effect in nanoferroics. *Physical Review B*, 79(16), 165433.

95. Eliseev, E. A., Morozovska, A. N., Glinchuk, M. D., Zaulychny, B. Y., Skorokhod, V. V. and Blinc, R. (2010). Surface-induced piezomagnetic, piezoelectric, and linear magnetoelectric effects in nanosystems. *Physical Review B*, 82(8), 085408.

96. Eringen, A. C. and Edelen, D. G. B. (1972). On nonlocal elasticity. *International Journal of Engineering Science*, 10(3), 233–248.

97. Erturk, A. and Inman, D. J. (2011). Broadband piezoelectric power generation on high-energy orbits of the bistable Duffing oscillator with electromechanical coupling. *Journal of Sound and Vibration*, 330, 2339–2353.

98. Erturk, A. and Inman, D. J. (2011). *Piezoelectric Energy Harvesting* (John Wiley & Sons, New York).

99. Esmaeili, M. and Tadi Beni, Y. (2019). Vibration and buckling analysis of functionally graded flexoelectric smart beam. *Journal of Applied and Computational Mechanics*, 5(5), 900–917.

100. Fan, K., Chang, J., Pedrycz, W., Liu, Z. and Zhu, Y. (2015). A nonlinear piezoelectric energy harvester for various mechanical motions. *Applied Physics Letters*, 106, 223902.

101. Fan, K., Tan, Q., Zhang, Y., Liu, S., Cai, M. and Zhu, Y. (2018). A monostable piezoelectric energy harvester for broadband low-level excitations. *Applied Physics Letters*, 112, 123901.

102. Fan, T. (2021). Energy harvesting from a nanopiezoelectric/piezomagnetic sandwich beam with porous properties. *Journal of Sandwich Structures & Materials*, 23(7), 3280–3302.

103. Fan, M., Yu, P. and Xiao, Z. (2022). An artificial neural network model for multi-flexoelectric actuation of plates. *International Journal of Smart and Nano Materials*, 13(4), 713–734.

104. Fatuzzo, E. (1960). Bias in ferroelectric colemanite. *Journal of Applied Physics*, 31, 1029–1034.

105. Fesenko, E. G., Smotrakov, V. G., Geguzina, G. A., Shuvaeva, E. T., Komarov, V. D., Gavrilyachenko, V. G. and Gagarina, E. S. (1994). New bismuth-containing layered perovskite-like oxides. *Inorganic Materials*, 30, 1446–1449.

106. Forbess, M. J., Seraji, S., Wu, Y., Nguyen, C. P. and Cao, G. Z. (2000). Dielectric properties of layered perovskite $Sr_{1-x}A_xBi_2Nb_2O_9$ ferroelectrics ($A =$ La, Ca and $x = 0.1$). *Applied Physics Letters*, 76, 2934–2936.

107. Foschini, C. R., Joshi, P. C., Varela, J. A. and Desu, S. B. (1999). Properties of $BaBi_2Ta_2O_9$ thin films prepared by chemical solution deposition technique for dynamic random-access memory applications. *Journal of Materials Research*, 14, 1860–1864.

108. Fouskova, A. and Cross L. E. (1970). Cross dielectric properties of bismuth. *Journal of Applied Physics*, 41, 2834–2838.

109. Fridkin, V. M. (1966). Some effects due to electron-phonon interaction during a phase transition in a ferroelectric-semiconductor. *JETF Letters*, 3, 252–255.

110. Friswell, M. I., Ali, S. F., Bilgen, O., Adhikari, S., Lees, A. W. and Litak, G. (2012). Non-linear piezoelectric vibration energy harvesting from a vertical cantilever beam with tip mass. *Journal of Intelligent Material Systems and Structures*, 23, 1505–1521.

111. Friswell, M. I., Bilgen, O., Ali, S. F., Litak, G. and Adhikari, S. (2015). The effect of noise on the response of a vertical cantilever beam energy harvester. *ZAMM.* 95, 433–443.

112. Galchev, T., Kim, H. and Najafi, K. (2011). Micro power generator for harvesting low-frequency and nonperiodic vibrations. *Journal of Microelectromechanical Systems*, 20, 852–866.

113. Gammaitoni, L., Vocca, H., Neri, I., Travasso, F. and Orfei, F. (2011). Vibration energy harvesting: Linear and nonlinear oscillator approaches. In *Sustainable Energy Harvesting Technologies–Past, Present and Future*, ed. Tan, Y. K. (InTech, Rijeka, Croatia) pp. 169–190.

114. Gaudenzi, P. (2009). *Smart Structures: Physical Behavior, Mathematical Modelling and Applications.* (John Wiley & Sons, USA).

115. Gebrael, T., Kanj, A., Farhat, D., Shehadeh, M. and Lakkis, I. (2020). Self sustained thermally induced gas-damped oscillations of bimetal cantilevers with application to the design of a new pyroelectric micro energy harvester. *Journal of Physics D: Applied Physics*, 53(19), 195501.

116. Geguzina, G. A., Shuvaev, A. T., Shuvaeva, E. T., Shilkina, L. A. and Vlasenko, V. G. (2005). Synthesis and structure of new phases of the type $A_{m-1}Bi_2B_mO_{3m+3}$ ($m = 2$). *Crystallography Reports*, 50, 59–64.

117. Geguzina, G. A., Shuvaev, A. T., Vlasenko, V. G., Shuvaeva, E. T. and Shilkina, L. A. (2003). Synthesis and structure of new phases of the type $Bi_2A_{m-1}B_mO_{3m+3}$ ($m = 3$). *Crystallography Reports*, 48, 406–412.

118. Gelfuso, M. V., Thomazini, D. and Eiras, J. A. (1999). Synthesis and structural, ferroelectric, and piezoelectric properties of $SrBi_4Ti_4O_{15}$ ceramics. *Journal of the American Ceramic Society*, 82, 2368–2372.

119. Ghemari, Z., Saad, S. and Khettab, K. (2019). Improvement of the vibratory diagnostic method by evolution of the piezoelectric sensor performances. *International Journal of Precision Engineering and Manufacturing*, 20(8), 1361–1369.

120. Ghobadi, A., Tadi Beni, Y. and Golestanian, H. (2020). Size dependent nonlinear bending analysis of a flexoelectric functionally graded nano-plate under thermo-electro-mechanical loads. *Journal of Solid Mechanics*, 12(1), 33–56.

121. Gibus, D., Gasnier, P., Morel, A., Formosa, F., Charleux, L., Boisseau, S. and Badel, A. (2020). Strongly coupled piezoelectric cantilevers for broadband vibration energy harvesting. *Applied Energy*, 277, 115518.

122. Gidde, R. R., Pawar, P. M. and Dhamgaye, V. P. (2020). Fully coupled modeling and design of a piezoelectric actuation based valveless micropump for drug delivery application. *Microsystem Technologies*, 26, 633–645.

123. Gilbarg, D. and Trudinger, N. S. (2001). *Elliptic Partial Differential Equations of Second Order* (Springer Berlin, Heidelberg).

124. Ginsburg, A. I. and Karapetyants, N. K. (1994). Fractional integro-differentiation in Hölder classes of variable order. *Doklady RAN*, 339(4), 439–441 (in Russian).

125. Ginzburg, V. L. (1950). Theory of ferroelectric phenomena. *Physics-Uspekhi (Advances in Physical Sciences)*, 38, 490–525.

126. Ginzburg, V. L. (1960). Several remarks on second-order phase transitions in the microscopic theory of ferroelectrics. *Physics of the Solid State*, 2, 2031–2043.

127. Gobat, G., Opreni, A., Fresca, S., Manzoni, A. and Frangi, A. (2022). Reduced order modeling of nonlinear microstructures through proper orthogonal decomposition. *Mechanical Systems and Signal Processing*, 171, 108864.

128. Goldschmidt, V. M. (1926). Die gesetze der krystallochemie. *Naturwissenschaften*, 14, 477–485.

129. Grudén, M., Hinnemo, M., Dancila, D., Zherdev, F., Edvinsson, N., Brunberg, K., Andersson, L., Byström, R. and Rydberg, A. (2014). Field operational testing for safety improvement of freight trains using wireless monitoring by sensor network. *IET Wireless Sensor Systems*, 4, 54–60.

130. Gu, Y., Liu, W., Zhao, C. and Wang, P. (2020). A goblet-like non-linear electromagnetic generator for planar multi-directional vibration energy harvesting. *Appl Energy*, 266, 114846.

131. Guha, S. and Singh, A. K. (2022). Frequency shifts and thermoelastic damping in distinct micro-/nano-scale piezothermoelastic fiber-reinforced composite beams under three heat conduction models. *Journal of Ocean Engineering and Science* (in press).

132. Guo, C. X., Guai, G. H. and Li, C. M. (2011). Graphene based materials: Enhancing solar energy harvesting. *Advanced Energy Materials*, 1, 448–452.

133. Gupta, J., Park, S. S., Bondy, B., Felner, E. I. and Prausnitz, M. R. (2011). Infusion pressure and pain during microneedle injection into skin of human subjects. *Biomaterials*, 32, 6823–6831.

134. Guseinov, A. I. and Mukhtarov, K. S. (1980). *Introduction to the Theory of Nonlinear Singular Equations* (Nauka, Moscow) (in Russian).

135. Gusev, A. A., Avvakumov, E. G., Isupov, V. P., Reznichenko, L. A., Verbenko, I. A., Miller, A. I. and Cherpakov, A. V. (2011). Mechanochemical synthesis of piezoelectrics on the base of lead zirconate titanate. In *Piezoelectric Materials and Devices*, ed. Parinov, I. A. (Nova Science Publishers, New York) pp. 189–234.

136. Guyomar, D. and Lallart, M. (2011). Recent progress in piezoelectric conversion and energy harvesting using nonlinear electronic interfaces and issues in small scale implementation. *Micromachines*, 2, 274–294.

137. Haldkar, R. K., Cherpakov, A. V. and Parinov, I. A. (2022). Modeling, Analysis and design optimizations of an axial-type piezoelectric energy generator for optimal output. *Smart Materials and Structures*, 31, 065019.

138. Haldkar, R. K., Cherpakov, A. V., Parinov, I. A. and Yakovlev, V. E. (2022). Comprehensive numerical analysis of a porous piezoelectric ceramic for axial load energy harvesting. *Applied Sciences*, 12, 10047–10060.

139. Haldkar, R. K., Gupta, V. K. and Sheorey, T. (2017). Modeling and flow analysis of piezoelectric based micropump with various shapes of microneedle. *Journal of Mechanical Science and Technology*, 31, 2933–2941.

140. Haldkar, R. K., Gupta, V. K., Sheorey, T. and Parinov, I. A. (2021). Design, modeling, and analysis of piezoelectric-actuated device for blood sampling. *Applied Sciences*, 11, 8449–8461.

141. Haldkar, R. K., Khalatkar, A., Gupta, V. K. and Sheorey, T. (2021). New piezoelectric actuator design for enhance the micropump flow. *Materials Today: Proceedings*, 44, 776–781.

142. Haldkar, R. K. and Parinov, I. A. (2021). Wind energy harvesting from artificial grass by using micro fibre composite. In *Physics and Mechanics of New Materials and Their Applications. Proc. Int. Conf. PHENMA 2020*, Vol. 10, eds. Parinov, I. A., Chang, S. H., Kim, Y. H. and Noda, N. A. (Springer, Cham, Switzerland) pp. 511–518.

143. Haldkar, R. K., Sheorey, T. and Gupta, V. K. (2018). The effect of operating frequency and needle diameter on performance of piezoelectric micropump. In *Advanced Materials — Proceedings of the International Conference on "Physics and Mechanics of New Materials*

and Their Applications", *PHENMA 2017*, Vol. 207, eds. Parinov, I., Chang, S. H. and Gupta, V. K. (Springer, Cham, Switzerland) pp. 567–578.

144. Haldkar, R. K., Sheorey, T., Gupta, V. K. and Ansari, M. Z. (2017). Four segment piezo based micropump. In *Smart Sensors, Actuators, and MEMS VIII*, Vol. 10246, eds. Fonseca, L., Prunnila, M., Peiner, E. (SPIE, Bellingham, Washington USA) p. 102461B.

145. Haluska, M. S. and Misture, S. T. (2004). Crystal structure refinements of the three-layer Aurivillius ceramics $Bi_2Sr_{2-x}A_xNb_2TiO_{12}$ (A = Ca; Ba; x = 0; 0:5; 1) using combined X-ray and neutron powder diffraction. *Journal of Solid State Chemistry*, 177, 1965–1975.

146. Hamdia, K. M., Ghasemi, H., Bazi, Y., AlHichri, H., Alajlan, N. and Rabczuk, T. (2019). A novel deep learning based method for the computational material design of flexoelectric nanostructures with topology optimization. *Finite Elements in Analysis and Design*, 165, 21–30.

147. Hardy, G. H. and Littlewood, J. E. (1928). Some properties of fractional integrals. I. *Mathematische Zeitschrift*, 17(4), 565–606.

148. Harne, R., Sun, A. and Wang, K. (2016). Leveraging nonlinear saturation-based phenomena in an L-shaped vibration energy harvesting system. *Journal of Sound and Vibration*, 363, 517–531.

149. Harne, R. L. and Wang, K. W. (2013). A review of the recent research on vibration energy harvesting via bistable systems. *Smart Materials and Structures*, 22, 023001.

150. Harris, P., Arafa, M., Litak, G., Bowen, C. R. and Iwaniec, J. (2017). Output response identification in a multistable system for piezoelectric energy harvesting. *European Physical Journal B*, 90, 20.

151. Hervoches, C. H., Irvine, J. T. S. and Lightfoot, P. (2001). Two high-temperature paraelectric phases in $Sr_{0.85}Bi_{2.1}Ta_2O_9$. *Physical Review B*, 64, 100102.

152. Hervoches, C. H. and Lightfoot, P. (2000). Cation disorder in three-layer Aurivillius phases: Structural studies of $Bi_{2-x}Sr_{2+x}Ti_{1-x}Nb_{2+x}O_{12}$ ($0 < x < 0.8$) and $Bi_{4-x}La_xTi_3O_{12}$ ($x = 1, 2$). *Journal of Solid State Chemistry*, 153, 66–73.

153. Hervoches, C. H., Snedden, A., Riggs, R., Kilcoyne, S. H., Manuel, P. and Lightfoot, P. (2002). Structural behavior of the four-layer Aurivillius-phase ferroelectrics $SrBi_4Ti_4O_{15}$ and $Bi_5Ti_3FeO_{15}$. *Journal of Solid State Chemistry*, 164, 280–291.

154. Hirose, M., Suzuki, T., Oka, H., Itakura, K., Miyauchi, Y. and Tsukada, T. (1999). Piezoelectric properties of $SrBi_4Ti_4O_{15}$-based ceramics. *Japan Journal of Applied Physics*, 38, 5561–5563.

155. Homayouni-Amlashi, A., Mohand-Ousaid, A. and Rakotondrabe, M. (2020). Analytical modelling and optimization of a piezoelectric cantilever energy harvester with in-span attachment. *Micromachines*, 11(6), 591.

156. Hosseini, S. M. H. and Beni, Y. T. (2023). Free vibration analysis of rotating piezoelectric/flexoelectric microbeams. *Applied Physics A*, 129(5), 330.

157. Huang, S., Qi, L., Huang, W., Shu, L., Zhou, S. and Jiang, X. (2018). Flexoelectricity in dielectrics: Materials, structures and characterizations. *Journal of Advanced Dielectrics*, 8(2), 1830002.

158. Huang, S. M., Feng, C. D., Chen, L. D. and Wen, X. W. (2005). Dielectric properties of $SrBi_{2-x}Pr_xNb_2O_9$ ceramics ($x = 0$, 0.04 and 0.2). *Solid State Communications*, 133, 375–379.

159. Huguet, T., Badel, A., Druet, O. and Lallart, M. (2018). Drastic bandwidth enhancement of bistable energy harvesters: Study of sub-harmonic behaviors and their stability robustness. *Applied Energy*, 226, 607–617.

160. Huguet, T., Badel, A. and Lallart, M. (2019). Parametric analysis for optimized piezoelectric bistable vibration energy harvesters. *Smart Materials and Structures*, 28(11), 115009.

161. Hung, P. T., Phung-Van, P. and Thai, C. H. (2023). Small scale thermal analysis of piezoelectric–piezomagnetic FG microplates using modified strain gradient theory. *International Journal of Mechanics and Materials in Design*, 1–23.

162. Hutson, V., Pym, J. S. and Cloud, M. J. (2005). *Applications of Functional Analysis and Operator Theory* (Elsevier, Amsterdam).

163. Hyatt, N. C., Hriljac, J. A. and Comyn, T. P. (2003). Cation disorder in $Bi_2Ln_2Ti_3O_{12}$ Aurivillius phases (Ln = La, Pr, Nd and Sm). *Materials Research Bulletin*, 38, 837–846.

164. Hyatt, N. C., Reaney, I. M. and Knight, K. S. (2005). Ferroelectric-paraelectric phase transition in the $n = 2$ Aurivillius phase $Bi_3Ti_{1.5}W_{0.5}O_9$: A neutron powder diffraction study. *Phycical Review B*, 71, 024119.

165. Ikegami, S. and Ueda, I. (1974). Piezoelectricity in ceramics of ferroelectric bismuth compounds with layer structure. *Japan Journal of Applied Physics*, 13, 1572–1577.

166. Iovane, G. and Nasedkin, A. V. (2019). Finite element modelling of ceramomatrix piezocomposites by using effective moduli method with different variants of boundary conditions. *Materials Physics and Mechanics*, 42, 1–13.

167. Islam, M. S., Lazure, S., Vannier, R.-N., Nowogrocki, G. and Mairesse, G. (1998). Structural and computational studies of Bi_2WO_6

based oxygen ion conductors. *Journal of Materials Chemistry*, 8, 655–660.

168. Ismailzade, I. G. (1960). X-ray study of the structure of some new ferroelectrics with a layered structure. *Izv. Academy of Sciences of the USSR*, 24, 1198–1202 (in Russian).

169. Ismailzade, I. G., Nesterenko, V. I., Mirishli, F. A. and Rustamov, P. G. (1967). X-ray and electrical studies of the $Bi_4Ti_3O_{12}$–$BiFeO_3$ system. *Crystallography Reports*, 12, 468–473.

170. Ismunandar, Hunter, B. A. and Kennedy, B. J. (1998). Cation disorder in the ferroelectric Aurivillius phase $PbBi_2Nb_2O_9$: An anamolous dispersion X-ray diffraction study. *Solid State Ionics*, 112, 281–289.

171. Ismunandar, Kamiyama, T., Hoshikawa, A., Zhou, Q., Kennedy, B. J., Kubota, Y. and Kato, K. (2004). Structural studies of five layer Aurivillius oxides: $A_2Bi_4Ti_5O_{18}$ (A = Ca, Sr, Ba and Pb). *Journal of Solid State Chemistry*, 177, 4188–4196.

172. Ismunandar, Kennedy, B. J., Gunawan and Marsongkohadi (1996). Structure $ABi_2Nb_2O_9$ (A = Sr, Ba): Refinement of powder neutron diffraction data. *Journal of Solid State Chemistry*, 126, 135–141.

173. Isupov, V. A. (1994). Properties of perovskite-like layered ferroelectric compounds of the $A_{T-1}B_2M_TO_{3T+3}$ type. *Russian Journal of Inorganic Chemistry*, 39, 731–737.

174. Isupov, V. A. (1996). Crystal chemical aspects of the bismuth-containing layered compounds of the $A_{m-1}Bi_2B_mO_{3m+3}$ type. *Ferroelectrics*, 189, 211–227.

175. Isupov, V. A. (1997). Curie temperatures of $A_{m-1}Bi_2M_mO_{3m+3}$ layered ferroelectrics. *Inorganic Materials*, 33, 936–940.

176. Iyer, S. and Venkatesh, T. A. (2014). Electromechanical response of (3-0,3-1) particulate, fibrous, and porous piezoelectric composites with anisotropic constituents: A model based on the homogenization method. *International Journal of Solids and Structures*, 51, 1221–1234.

177. James, A. R., Kumar, G. S., Bhimsankaram, T. and Suryanarayana, S. V. (1994). Studies on electrical conduction in $Bi_4SrTi4O_{15}$. *Bulletin of Materials Science*, 17, 951–958.

178. James, A. R., Kumar, G. S., Bhimsankaram, T. and Suryanarayana, S. V. (1996). Impedance Spectroscopic Studies in $SrBi_5FeTi_4O_{18}$. *Ferroelectrics*, 189, 81–90.

179. James, A. R., Kumar, G. S., Kumar, M., Suryanarayana, S. V. and Bhimasankaram, T. (1997). Magnetic and magnetoelectric studies in polycrystalline $LaBi_4FeTi_3O_{15}$. *Modern Physics Letters B*, 11, 633–644.

180. Jartych, E., Mazurek, M., Lisińka-Czekaj, A. and Czekaj, D. (2010). Hyperfine interactions in some Aurivillius $Bi_{m+1}Ti_3Fe_{m-3}O_{3m+3}$ compound. *Journal of Magnetism and Magnetic Materials*, 322, 51–55.

181. Jia, Y., Wei, X., Xu, L., Wang, C., Lian, P., Xue, S. and Shi, Y. (2019). Multiphysics vibration FE model of piezoelectric macro fibre composite on carbon fibre composite structures. *Composites Part B: Engineering*, 161, 376–385.

182. Jiang, J., Liu, S., Feng, L. and Zhao, D. (2021). A review of piezoelectric vibration energy harvesting with magnetic coupling based on different structural characteristics. *Micromachines*, 12, 436.

183. Jiménez, B., Pardo, L., Castro, A., Millán, P., Jiménez, R., Elaatmani, M. and Oualla, M. (2000). Influence of the preparation on the microstructure and ferroelectricity of the $(SBN)_{1-x}(BTN)_x$ ceramics. *Ferroelectrics*, 241, 279–286.

184. Jin, Q., Ren, Y., Jiang, H. and Li, L. (2021). A higher-order size-dependent beam model for nonlinear mechanics of fluid-conveying FG nanotubes incorporating surface energy. *Composite Structures*, 269, 114022.

185. Jung, S. M. and Yun, K. S. (2010). Energy-harvesting device with mechanical frequency-up conversion for increased power efficiency and wideband operation. *Applied Physics Letters*, 96, 111906.

186. Kajewski, D., Ujma, Z., Szot, K. and Paweczyk, M. (2009). Dielectric properties and phase transition in $SrBi_2Nb_2O_9$–$SrBi_2Ta_2O_9$ solid solution. *Ceramics International*, 35, 2351–2355.

187. Kang, M. and Yeatman, E. M. (2020). Coupling of piezo-and pyro-electric effects in miniature thermal energy harvesters. *Applied Energy*, 262, 114496.

188. Karami, M. A., Farmer, J. R. and Inman D. J. (2013). Parametrically excited nonlinear piezoelectric compact wind turbine. *Renewable Energy*, 50, 977–987.

189. Karami, M. A., Inman, D. J. (2012). Powering pacemakers from heartbeat vibrations using linear and nonlinear energy harvesters. *Applied Physics Letters*, 100, 042901.

190. Karapetyants, N. K. and Ginsburg, A. I. (1995). Fractional integro-differentiation in Hölder classes of arbitrary order. *Georgian Mathematical Journal*, 2(2), 141–150.

191. Kawun, P., Leahy, S. and Lai, Y. (2016). A thin PDMS nozzle/diffuser micropump for biomedical applications. *Sensors & Actuators: A. Physical*, 249, 149–154.

192. Kennedy, B. J., Kubota, Y., Hunter, B., Ismunandar and Kato, K. (2003). Structural phase transitions in the layered bismuth oxide BaBi$_4$Ti$_4$O$_{15}$. *Solid State Communication*, 126, 653–658.

193. Keve, E. T., Bye, K. L., Whipps, P. W. and Annis, A. D. (1971). Structural inhibition of ferroelectric switching in triglycine sulphate. *Ferroelectrics*, 3, 39–48.

194. Khalatkar, A., Gupta, V. K. and Haldkar, R. (2011). Modeling and simulation of cantilever beam for optimal placement of piezoelectric actuators for maximum energy harvesting. *Smart Nano-Micro Materials and Devices*, 8204, 82042G.

195. Khalatkar, A. M., Haldkar, R. H. and Gupta, V. K. (2011). Finite element analysis of cantilever beam for optimal placement of piezoelectric actuator. In *Applied Mechanics and Materials* (Trans Tech Publications, Ltd., Stafa-Zurich, Switzerland) Vol. 110–116, pp. 4212–4219.

196. Khalatkar, A. M., Kumar, R., Haldkar, R. and Jhodkar D. (2019). Arduino-based tuned electromagnetic shaker using relay for MEMS cantilever beam. In *Smart Technologies for Energy, Environment and Sustainable Development. Lecture Notes on Multidisciplinary Industrial Engineering* (Springer Nature, Singapore), pp. 795–801.

197. Khasbulatov, S., Cherpakov, A., Parinov, I., Andryushin, K., Shlkina, L., Aleshin, V., Andryushina, I., Mardaliev, B., Gordienko, D., Verbenko, I. and Reznichenko, L. (2020). Destruction phenomena in ferroactive materials. *Journal of Advanced Dielectrics*, 10, 2050012.

198. Khazaee, M., Rezania, A. and Rosendahl, L. (2022). Piezoelectric resonator design and analysis from stochastic car vibration using an experimentally validated finite element with viscous-structural damping model. *Sustainable Energy Technologies and Assessments*, 52, 102228.

199. Khorasani, M., Soleimani-Javid, Z., Arshid, E., Amir, S. and Civalek, Ö. (2021). Vibration analysis of graphene nanoplatelets' reinforced composite plates integrated by piezo-electromagnetic patches on the piezo-electromagnetic media. *Waves in Random and Complex Media*, 1–31.

200. Kikuchi, T., Watanabe, A. and Uchida, K. (1977). A family of mixed layer type bismuth compounds. *Materials Research Bulletin*, 12, 299–304.

201. Kilbas, A. A., Srivastava, H. M. and Trujillo, J. J. (2006). *Theory and Applications of Fractional Differential Equations* (Elsevier, Amsterdam).

202. Kim, H. S., Kim, J. H. and Kim, J. (2011). A review of piezoelectric energy harvesting based on vibration. *International Journal of Precision Engineering and Manufacturing*, 12, 1129–1141.

203. Kim, J. E., Lee, S. and Kim, Y. Y. (2019). Mathematical model development, experimental validation and design parameter study of a folded two-degree-of-freedom piezoelectric vibration energy harvester. *International Journal of Precision Engineering and Manufacturing-Green Technology*, 6, 893–906.

204. Kishore, R. A., Vučković, D. and Priya, S. (2014). Ultra-low wind speed piezoelectric windmill. *Ferroelectrics*, 460, 98–107.

205. Klyushnichenko, V. A. and Kramarov, Y. A. (1974). Synthesis of a non-uniformly polarized piezoelectric transducer. In *Acoustic Methods and Resources of Ocean Research* (Far Eastern State University Press, Vladivostok) pp. 71–72 (in Russian).

206. Komarov, A. V., Geguzina, G. A., Gagarina, E. S., Komarov, V. D., Leiderman, A. V., Shuvaeva, E. T., Shuvaev, A. T. and Fesenko, E. G. (2000). Synthesis and ferroelectric properties of layered perovskite-like oxides $A^{II}Bi_3Ti_2NbO_{12}$, (A^{II} = Sr, Pb). *Inorganic Materials*, 36, 237–242.

207. Kouritem, S. A., Al-Moghazy, M. A., Noori, M. and Altabey, W. A. (2022). Mass tuning technique for a broadband piezoelectric energy harvester array. *Mechanical Systems and Signal Processing*, 181, 109500.

208. Krasilnikov, V. A. and Krylov, V. V. (1984). *Introduction to Physical Acoustics* (Science, Moscow), (in Russian).

209. Kubel, F. and Schmid, H. (1992). X-ray room temperature structure from single crystal data, powder diffraction measurements and optical studies of the Aurivillius phase $Bi_5(Ti_3Fe)O_{15}$. *Ferroelectrics*, 129, 101–112.

210. Kumar, A., Ali, S. F. and Arockiarajan, A. (2018). Exploring the benefits of an asymmetric monostable potential function in broadband vibration energy harvesting. *Applied Physics Letters*, 112, 233901.

211. Kumari, N. (2021). Contribution to the design and development of hybrid thermal-vibrational piezoelectric energy harvester (Doctoral dissertation, Université Bourgogne Franche-Comté, France).

212. Kurths, J., Boccaletti, S., Grebogi, C. and Lai Y. C. (2003). Introduction: Control and synchronization in chaotic dynamical systems. *Chaos*, 13, 126–127.

213. Kwak, W. and Lee, Y. (2021). Optimal design and experimental verification of piezoelectric energy harvester with fractal structure. *Applied Energy*, 282, 116121.

214. Kwuimy, C. A. K., Litak, G. and Nataraj, C. (2015). Nonlinear analysis of energy harvesting systems with fractional order physical properties. *Nonlinear Dynamics*, 80, 491–501.
215. Lach, J., Wróbel, K., Wróbel, J. and Czerwínski, A. (2021). Applications of carbon in rechargeable electrochemical power sources: A review. *Energies*, 14, 2649.
216. Lallart, M., Anton, S. R. and Inman, D. J. (2010). Frequency self-tuning scheme for broadband vibration energy harvesting. *Journal of Intelligent Material Systems and Structures*, 21, 897–906.
217. Lallart, M., Yan, L., Miki, H., Sebald, G., Diguet, G., Ohtsuka, M. and Kohl, M. (2021). Heusler alloy-based heat engine using pyroelectric conversion for small-scale thermal energy harvesting. *Applied Energy*, 288, 116617.
218. Lallart, M., Zhou, S., Yang, Z., Yan, L., Li, K. and Chen, Y. (2020). Coupling mechanical and electrical nonlinearities: The effect of synchronized discharging on tristable energy harvesters. *Applied Energy*, 266, 114516.
219. Lan, C., Tang, L. and Qin, W. (2017). Obtaining high-energy responses of nonlinear piezoelectric energy harvester by voltage impulse perturbations. *The European Physical Journal Applied Physic*, 79, 20902.
220. Lan, M., Yang, W., Liang, X., Hu, S. and Shen, S. (2022). Vibration modes of flexoelectric circular plate. *Acta Mechanica Sinica*, 38(12), 422063.
221. Landau, L. D. (1969). On the theory of phase transitions. In *Landau L. D. Collection of Works*, Vol. 1 (Science, Moscow), pp. 234–252 (in Russian).
222. Landau, L. D. and Lifshits, E. M. (1978). *Statistical Physics*. 2nd ed. (Science, Moscow) (in Russian).
223. Landis, C. M. (2002). Fully coupled, multi-axial, symmetric constitutive laws for polycrystalline ferroelectric ceramics. *Journal of the Mechanics and Physics of Solids*, 50, 127–152.
224. Landkof, N. S. (1973). *Foundations of Modern Potential Theory* (Springer, Providence, RI).
225. Lang, S. B. (1974). *Sourcebook of Pyroelectricity*, Vol. 2 (CRC Press, USA).
226. Lang, S. B. and Muensit, S. (2006). Review of some lesser-known applications of piezoelectric and pyroelectric polymers. *Applied Physics A*, 85, 125–134.
227. Leadenham, S. and Erturk, A. (2014). M-shaped asymmetric nonlinear oscillator for broadband vibration energy harvesting: Harmonic balance analysis and experimental validation. *Journal of Sound and Vibration*, 333, 6209–6223.

228. Leadenham, S. and Erturk, A. (2015). Nonlinear M-shaped broadband piezoelectric energy harvester for very low base accelerations: Primary and secondary resonances. *Smart Materials and Structures*, 24, 055021.

229. Legusha, F. F. and Popov, Y. N. (2021). Acoustic wave absorption in a waveguide with impedance boundary conditions. *Transactions of the Krylov State Research Centre*, 2(396). 113–121 (in Russian).

230. Levanyuk, L. I. and Sannikov, D. G. (1974). Improper ferroelectrics. *Physics-Uspekhi (Advances in Physical Sciences)*, 112, 561–589.

231. Levanyuk, L. I. and Sannikov, D. G. (1976). Theory of phase transitions in ferroelectrics with the formation of a superstructure that is not a multiple of the initial parameter. *Physics of the Solid State*, 18, 423–428.

232. Li, C. G., Lee, C. Y., Lee, K. and Jung, H. (2013). An optimized hollow microneedle for minimally invasive blood extraction. *Biomedical Microdevices*, 15, 17–25.

233. Li, C. G., Lee, K., Lee, C. Y., Dangol, M. and Jung, H. (2012). A minimally invasive blood-extraction system: Elastic self-recovery actuator integrated with an ultrahigh-aspect-ratio microneedle. *Advanced Materials*, 24, 4583–4586.

234. Li, H., Ding, H. and Chen, L. (2019). Chaos threshold of a multistable piezoelectric energy harvester subjected to wake-galloping. *International Journal of Bifurcation and Chaos*, 29, 1950162.

235. Li, J., He, X., Yang, X. and Liu, Y. (2020). A consistent geometrically nonlinear model of cantilevered piezoelectric vibration energy harvesters. *Journal of Sound and Vibration*, 486, 115614.

236. Li, J.-B., Huang, Y. P., Rao, G. H., Liu, G. Y., Luo, J., Chen, J. R. and Liang, J. K. (2010). Ferroelectric transition of Aurivillius compounds $Bi_5Ti_3FeO_{15}$ and $Bi_6Ti_3Fe_2O_{18}$. *Applied Physics Letters*, 96, 222903.

237. Li, X., Guo, M. and Dong, S. (2011). A flex-compressive-mode piezoelectric transducer for mechanical vibration/strain energy harvesting. *IEEE Transactions of Ultrasonics Ferroelectrics and Frequency Control*, 58, 698–703.

238. Li, Z., Xin, C., Peng, Y., Wang, M., Luo, J., Xie, S. and Pu, H. (2021). Power density improvement of piezoelectric energy harvesters via a novel hybridization scheme with electromagnetic transduction. *Micromachines*, 12, 803.

239. Liang, H., Hao, G. and Olszewski, O. Z. (2021). A review on vibration-based piezoelectric energy harvesting from the aspect of compliant mechanisms. *Sensors and Actuators, A: Physical*, 331, 112743.

240. Liao, Y. and Liang, J. (2019). Unified modeling, analysis and comparison of piezoelectric vibration energy harvesters. *Mechanical Systems and Signal Processing*, 123, 403–425.
241. Lin, G.-J., Wang, H.-P., Lien, D.-H., Fu, P.-H., Chang, H.-C., Ho, C.-H., Lin, C.-A., Lai, K.-Y. and He, J.-H. (2014). A broadband and omnidirectional light-harvesting scheme employing nanospheres on Si solar cells. *Nano Energy*, 6, 36–43.
242. Lin, J. T. and Alphenaar, B. (2010). Enhancement of energy harvested from a random vibration source by magnetic coupling of a piezoelectric cantilever. *Journal of Intelligent Material Systems and Structures*, 21, 1337–1341.
243. Lisińska-Czekaj, A., Czekaj, D., Surowiak, Z., Ilczuk, J., Plewa, J., Leyderman, A. V., Gagarina, E. S., Shuvaev, A. T. and Fesenko, E. G. (2004). Synthesis and dielectric properties of $A_{m-1}Bi_2B_mO_{3m+3}$ ceramic ferroelectrics with $m = 1.5$. *Journal of the European Ceramic Society*, 24, 947–951.
244. Litak, G., Friswell, M. I. and Adhikari, S. (2010). Magnetopiezoelastic energy harvesting driven by random excitations. *Applied Physics Letters*, 96, 214103.
245. Litak, G., Friswell, M. I., Kwuimy, C. A. K., Adhikari, S. and Borowiec, M. (2012). Energy harvesting by two magnetopiezoelastic oscillators with mistuning. *Theoretical and Applied Mechanics Letters*, 2, 043009.
246. Liu, D., Al-Haik, M. Y., Zakaria, M. Y. and Hajj, M. R. (2018). Piezoelectric energy harvesting using L-shaped structures. *Journal of Intelligent Material Systems and Structures*, 29, 1206–1215.
247. Liu, H., Zhong, J., Lee, C., Lee, S.-W. and Lin, L. (2018). A comprehensive review on piezoelectric energy harvesting technology: Materials, mechanisms, and applications. *Applied Physics Reviews*, 5, 041306.
248. Liu, J. and Li, Q. (2021). Coupled mode sound propagation in inhomogeneous stratified waveguides. *Applied Sciences*, 11, 3957.
249. Liu, L. and Yuan, F. G. (2011). Nonlinear vibration energy harvester using diamagnetic levitation. *Applied Physics Letters*, 98, 203507.
250. Lomanova, N. A. and Gusarov, V. V. (2011). On the limiting thickness of a perovskite-like block in Aurivillius phases in the Bi_2O_3–Fe_2O_3–TiO_2 system. *Nanosystems: Physics, Chemistry, Mathematics*, 2, 93–101.
251. Lomanova, N. A., Morozov, M. I., Ugolkov, V. L. and Gusarov, V. V. (2006). Impedance spectroscopy of polycrystalline materials based on Aurivillius phases of the $Bi_4Ti_3O_{12}$–$BiFeO_3$ system. *Inorganic Materials*, 42(2), 1–7.

252. Ma, Z. and Arvin, H. (2023). Nonlinear thermo-electro-mechanical free vibrations of sandwich nanocomposite beams bonded with sensor layers considering pyroelectricity. *Engineering Analysis with Boundary Elements*, 148, 90–103.

253. Machado, J. A. T., Silva, M. F., Barbosa, R. S., Jesus, I. S., Reis, C. M., Marcos, M. G. and Galhano, A. F. (2010). Some applications of fractional calculus in engineering. *Mathematical Problems in Engineering*, 2010, 639801.

254. Macquart, R., Kennedy, B. J., Hunter, B. A. and Howard, C. J. (2002). High-temperature structural studies of $PbBi_2M_2O_9$ (M = Nb and Ta). *Journal of Physics: Condensed Matter*, 14, 7955–7960.

255. Macquart, R., Kennedy, B. J., Hunter, B. A., Howard, C. J. and Shimakawa, Y. (2002). Structural phase transitions in the ferroelectric oxide $SrBi_2Ta_2O_9$. *Integrated Ferroelectrics*, 44, 101–112.

256. Malikan, M. and Eremeyev, V. A. (2020). Free vibration of flexomagnetic nanostructured tubes based on stress-driven nonlocal elasticity. *Analysis of Shells, Plates, and Beams: A State of the Art Report*, 215–226.

257. Malikan, M. and Eremeyev, V. A. (2020). On the dynamics of a visco–piezo–flexoelectric nanobeam. *Symmetry*, 12(4), 643.

258. Malikan, M. and Eremeyev, V. A. (2021). Flexomagnetic response of buckled piezomagnetic composite nanoplates. *Composite Structures*, 267, 113932.

259. Malikan, M. and Eremeyev, V. A. (2022). The effect of shear deformations' rotary inertia on the vibrating response of multi-physic composite beam-like actuators. *Composite Structures*, 297, 115951.

260. Malikan, M. and Eremeyev, V. A. (2023). On dynamic modeling of piezomagnetic/flexomagnetic microstructures based on Lord–Shulman thermoelastic model. *Archive of Applied Mechanics*, 93(1), 181–196.

261. Malikan, M., Eremeyev, V. A. and Żur, K. K. (2020). Effect of axial porosities on flexomagnetic response of in-plane compressed piezomagnetic nanobeams. *Symmetry*, 12(12), 1935.

262. Malikan, M., Uglov, N. S. and Eremeyev, V. A. (2020). On instabilities and post-buckling of piezomagnetic and flexomagnetic nanostructures. *International Journal of Engineering Science*, 157, 103395.

263. Mandal, T. K., Augustine, S., Gopalakrishnan, J. and Boullay, P. (2005). $Bi_4LnNb_3O_{15}$ (Ln = La, Pr, Nd) and $Bi_4LaTa_3O_{15}$: New intergrowth Aurivillius related phases. *Materials Research Bulletin*, 40, 920–927.

264. Mandal, T. K., Sivakumar, T., Augustine, S. and Gopalakrishnan, J. (2005). Heterovalent cation-substituted Aurivillius phases Bi_2SrNa

Nb_2TaO_{12} and $Bi_2Sr_2Nb_{3-x}M_xO_{12}$ (M = Zr, Hf, Fe, Zn). *Materials Science and Engineering B*, 121, 112–119.

265. Mangalasseri, A. S., Mahesh, V., Mukunda, S., Mahesh, V., Ponnusami, S. A. and Harursampath, D. (2022). Vibration-based energy harvesting characteristics of functionally graded magneto-electro-elastic beam structures using lumped parameter model. *Journal of Vibration Engineering & Technologies*, 10(5), 1705–1720.

266. Mann, B. P. and Owens, B. A. (2010). Investigations of a nonlinear energy harvester with a bistable potential well. *Journal of Sound and Vibration*, 329, 1215–1226.

267. Mao, X., Wang, W. and Chen, X. (2008). Electrical and magnetic properties of $Bi_5FeTi_3O_{15}$ compound prepared by inserting $BiFeO_3$ into $Bi_4Ti_3O_{12}$. *Solid State Communications*, 147, pp.186–189.

268. Mao, X., Wang, W., Chen, X., Lu, Y. (2009). Multiferroic properties of layer-structured $Bi_5Fe_{0.5}Co_{0.5}Ti_3O_{15}$ ceramics. *Applied Physics Letters*, 95, 082901–0829013.

269. Martínez-Sarrión, M.-L., Mestres, L., Herraiz, M., Balagurov, A. M., Beskrovniy, A. I., Vasilovskij, S. G. and Smirnov, L. S. (2002). Synthesis and characterization of a new aurivillius phase. *European Journal of Inorganic Chemistry*, 7, 1801–1805.

270. Maslovskaya, A. G. and Barabash, T. K. (2013). Dynamic simulation of polarization reversal processes in ferroelectric crystals under electron beam irradiation. *Ferroelectrics*, 442, 18–26.

271. Masoumi, A., Amiri, A. and Talebitooti, R. (2019). Flexoelectric effects on wave propagation responses of piezoelectric nanobeams via nonlocal strain gradient higher order beam model. *Materials Research Express*, 6(10), 1050d5.

272. Masuda, A. and Senda, A. (2012). Stabilization of a wide-band nonlinear vibration energy harvester by using a nonlinear self-excitation circuit. *Proceedings of SPIE*, 8341, 83411B.

273. Mazurek, M., Jartych, E., Lisińka-Czekaj, A., Czekaj, D. and Oleszak, D. (1994). Structure and hyperfine interactions of $Bi_9Ti_3Fe_5O_{27}$ multiferroic ceramic prepared by sintering and mechanical alloying methods. *Journal of Non-Crystalline Solids*, 356, 1997.

274. McCabe, E. E. and Greaves, C. (2005). Structural and magnetic characterisation of $Bi_2Sr_{1.4}La_{0.6}Nb_2MnO_{12}$ and its relationship to $Bi_2Sr_2Nb_2MnO_{12}$. *Journal of Materials Chemistry*, 15, 177–182.

275. Mercier, J. F. and Maurel, A. (2013). Acoustic propagation in nonuniform waveguides: Revisiting Webster equation using evanescent boundary modes. *Proceedings of the Royal Society A*, 469 (2156), 20130186.

276. Mikhlin, S. G. and Prössdorf, S. (1986). *Singular Integral Operators* (Springer-Verlag, Berlin).

277. Momeni-Khabisi, H. and Tahani, M. (2023). Coupled thermal stability analysis of piezomagnetic nano-sensors and nano-actuators considering the flexomagnetic effect. *European Journal of Mechanics-A/Solids*, 97, p.104773.

278. Montel, P. (1918). Sur les polynomes d'approximation. *Bulletin de la Société Mathématique de France*, 46, 151–196 (in French).

279. Moon, S.-Y., Choi, K. S., Jung, R. W., Lee, H. and Jung, D. (2002). Ferroelectric properties of substituted Aurivillius phases $SrBi_2Nb_{2-x}M_xO_9$ (M = Cr, Mo). *Bulletin of the Korean Chemical Society*, 23, 1463–1466.

280. Morangueira, Y. L. and Pereira, J. C. D. C. (2020). Energy harvesting assessment with a coupled full car and piezoelectric model. *Energy*, 210, 118668.

281. Moroz, L. I. and Maslovskaya, A. G. (2019). Hybrid stochastic fractal-based approach to modelling ferroelectrics switching kinetics in injection mode. *Matematicheskoe Modelirovanie*, 31(9), 131–144 (in Russian).

282. Moure, A. and Pardo, L. (2005). Microstructure and texture dependence of the dielectric anomalies and dc conductivity of Bi_3TiNbO_9 ferroelectric ceramics. *Journal of Applied Physics*, 97, 084103.

283. Moure, C., Fernandez, J. F., Villegas, M. and Duran, P. (1995). Processing and sintering of $CaBi_4Ti_4O_{15}$ powders for high temperature piezoelectric materials. *Euroceramics IV*, 5, 139–144.

284. Naderi, A., Quoc-Thai, T., Zhuang, X. and Jiang, X. (2023). Vibration analysis of a unimorph nanobeam with a dielectric layer of both flexoelectricity and piezoelectricity. *Materials*, 16(9), 3485.

285. Nan, C. W. (1994). Magnetoelectric effect in composites of piezoelectric and piezomagnetic phases. *Physical Review B*, 50(9), 6082.

286. Narita, F. and Fox, M. (2018). A review on piezoelectric, magnetostrictive, and magnetoelectric materials and device technologies for energy harvesting applications. *Advanced Engineering Materials* 20, 1700743.

287. Narolia, T., Gupta, V. K. and Parinov, I. A. (2020). Design and analysis of a shear mode piezoelectric energy harvester for rotational motion system. *Journal of Advanced Dielectrics*. 10, 7–10.

288. Nasedkin, A. V. (2011). *Finite-element Modeling of Piezoelectric Generators from Highly Porous Piezoceramics* (IGM NAN Ukraine, Kiev).

289. Nasedkin, A. V. (2015). Finite element design of piezoelectric and magnetoelectric composites with use of symmetric quasidefinite matrices. In *Advanced Materials — Studies and Applications*, eds. Parinov, I. A., Chang, S. H. and Theerakulpisut, S. (Nova Science Publishers, New York) pp. 109–124.

290. Nasedkin, A. V., Oganesyan, P. A. and Soloviev, A. N. (2021). Analysis of Rosen type energy harvesting devices from porous piezoceramics with great longitudinal piezomodulus. *ZAMM Zeitschrift fur Angewandte Mathematik und Mechanik*, 101, e202000129.
291. Naseer, R., Dai, H., Abdelkefi, A. and Wang, L. (2017). Piezomagnetoelastic energy harvesting from vortex-induced vibrations using monostable characteristics. *Appl. Energy*, 203, 142–153.
292. Naseer, R., Dai, H., Abdelkefi, A. and Wang, L. (2019). Comparative study of piezoelectric vortex-induced vibration-based energy harvesters with multi-stability characteristics. *Energies*, 13, 13010071.
293. Ngueuteu, G. M. and Woafo, P. (2012). Dynamics and synchronization analysis of coupled fractional-order nonlinear electromechanical systems. *Mech. Res. Commun.* 46, 20–25.
294. Nikolsky, S. M. and Lizorkin, P. I. (1984). Approximation by spherical polynomials. *Proceedings of MIAN SSSR*, 166, 186–200.
295. Noguchi, Y., Miyayama, M., Oikawa, K. and Kamiyama, T. (2004). Cation-vacancy- induced low coercive field in La-Modified $SrBi_2Ta_2O_9$. *Journal of Applied Physics*, 95, 4261–4266.
296. Noguchi, Y., Miyayama, M., Oikawa, K., Kamiyama, T., Osada, M. and Kakihana, M. (2002). Defect engineering for control of polarization properties in $SrBi_2Ta_2O_9$. *Japan Journal of Applied Physics*, 41, 7062–7075.
297. Nurul Adina, A. (2021). Design and simulation of hybrid piezopyroelectric energy conversion device (Doctoral dissertation, Universiti Malaya, Malaysia).
298. Odzijewicz, T., Malinowska, A. B. and Torres D. F. M. (2013). Fractional variational calculus of variable order. In *Advances in Harmonic Analysis and Operator Theory: The Stefan Samko Anniversary Volume*. eds. Almeida, A., Castro, L. and Speck, F.-O. (Birkhäuser, Basel, Switzerland), pp. 291–301.
299. Orrego, S., Shoele, K., Ruas, A., Doran, K., Caggiano, B., Mittal, R. and Kang, S. H. (2017). Harvesting ambient wind energy with an inverted piezoelectric flag. *Applied Energy*, 194, 212–222.
300. Panich, A. A., Marakhovskii, M. A. and Motin D. V. (2011). Crystal and ceramic piezoelectric. *Engineering Journal of Don*, 1 (in Russian). http://www.ivdon.ru/magazine/archive/n1y2011/325.
301. Parinov, I. A. (2013). *Microstructure and Properties of High-Temperature Superconductors*, (2nd ed.) (Springer, Heidelberg, New York, Dordrecht, London).
302. Parinov, I. A. (ed.) (2010). *Piezoceramic Materials and Devices* (Nova Science Publishers, New York).

303. Parinov, I. A. (ed.) (2012). *Ferroelectrics and Superconductors: Properties and Applications* (Nova Science Publishers, New York).
304. Parinov, I. A. (ed.). (2012). *Piezoelectric Materials and Devices* (Nova Science Publishers, New York).
305. Parinov, I. A. (ed.). (2012). *Piezoelectrics and Related Materials: Investigations and Applications* (Nova Science Publishers, New York).
306. Parinov, I. A. (ed.). (2013). *Nano- and Piezoelectric Technologies, Materials and Devices* (Nova Science Publishers, New York).
307. Parinov, I. A. (ed.). (2014). *Advanced Nano- and Piezoelectric Materials and their Applications* (Nova Science Publishers, New York).
308. Parinov, I. A. (ed.). (2015). *Piezoelectrics and Nanomaterials: Fundamentals, Developments and Applications* (Nova Science Publishers, New York).
309. Parinov, I. A. and Chang, S.-H. (eds.). (2013). *Physics and Mechanics of New Materials and Their Applications* (Nova Science Publishers, New York).
310. Parinov, I. A., Chang, S.-H. and Gupta, V. K. (eds.). (2018). *Proceedings of the International Conference on "Physics and Mechanics of New Materials and Their Applications", PHENMA 2017*, Vol. 207 (Springer Cham, Heidelberg, New York, Dordrecht, London).
311. Parinov, I. A., Chang, S.-H. and Jani, M. A. (eds.). (2017). *Advanced Materials–Techniques, Physics, Mechanics and Applications*, Vol. 193 (Springer Cham, Heidelberg, New York, Dordrecht, London).
312. Parinov, I. A., Chang, S.-H. and Kim, Y.-H. (eds.). (2019). *Advanced Materials — Proceedings of the International Conference on "Physics and Mechanics of New Materials and Their Applications", PHENMA 2018*, Vol. 224 (Springer Cham, Heidelberg, New York, Dordrecht, London).
313. Parinov, I. A., Chang, S.-H., Kim, Y.-H. and Noda N.-A. (eds.). (2021). *Proceedings of the International Conference on "Physics and Mechanics of New Materials and Their Applications, PHENMA 2020*, Vol. 10 (Springer Nature, Cham, Switzerland).
314. Parinov, I. A., Chang, S.-H. and Long B. T. (eds.). (2020). *Advanced Materials — Proceedings of the International Conference on "Physics and Mechanics of New Materials and Their Applications", PHENMA 2019*, Vol. 6 (Springer Nature, Cham, Switzerland).
315. Parinov, I. A., Chang, S.-H. and Soloviev, A. N. (eds.). (2023). *Proceedings of the International Conference on "Physics and Mechanics of New Materials and Their Applications", PHENMA 2021–2022*, Vol. 20 (Springer Nature, Cham, Switzerland).

316. Parinov, I. A., Chang, S.-H. and Soloviev, A. N. (eds.). (2023). *Phys. Mech. New Mater. Their Appl., 2021–2022* (Nova Science Publishers, New York).

317. Parinov, I. A., Chang, S.-H. and Theerakulpisut, S. (eds.). (2015). *Advanced Materials — Studies and Applications* (Nova Science Publishers, New York).

318. Parinov, I. A., Chang, S.-H. and Topolov, V. Y. (eds.). (2016). *Advanced Materials — Manufacturing, Physics, Mechanics and Applications*, Vol. 175 (Springer Cham, Heidelberg, New York, Dordrecht, London).

319. Parinov, I. A. and Cherpakov, A. V. (2022). Overview: State-of-the-art of energy harvesting based on piezoelectric devices for last decade. *Symmetry*, 14, 765–813.

320. Parinov, I. A., Cherpakov, A. V., Rozhkov, E. V., Soloviev, A. N. and Chebanenko, V. A. (2018). Program-signal generator "Sgenerator". *Russian Certificate of State Registration of a Computer Program*, No. RU 2018610408, 10.01.2018 (in Russian).

321. Parinov, I. A., Soloviev, A. N. and Cherpakov, A. V. (2016). The program "Vibrograf" for registration, visualization and processing of vibrations of structures. *Russian Certificate of State Registration of a Computer Program*, No. RU 2016612309, 24.02.2016 (in Russian).

322. Paulo, J. and Gaspar, P. (2010). *Proceedings of the World Congress on Engineering 2010 (WCE 2010), London, UK, 30 June–2 July 2010*, "Review and future trend of energy harvesting methods for portable medical devices." (London, UK), II.

323. Pellegrini, S. P., Tolou, N., Schenk, M. and Herder, J. L. (2013). Bistable vibration energy harvesters: A review. *Journal of Intelligent Material Systems and Structures*, 24, 1303–1312.

324. Petrov, A. and Rumyantseva, V. (2018). The acoustic waves propagation in a cylindrical waveguide with the laminar flow. *MATEC Web of Conferences*, 211, 04005.

325. Pham, T. D., Tarlakovskii, D. V. and Paimushin, V. N. (2021). Non-stationary bending of a finite electromagnetoelastic rod. *ZAMM-Journal of Applied Mathematics and Mechanics/Zeitschrift für Angewandte Mathematik und Mechanik*, 101(9), e202000316.

326. Polinger, V. Z. (2013). Ferroelectric phase transitions in cubic perovskites. *Journal of Physics: Conference Series*, 428, 012026.

327. Polyakova, T. V., Cherpakov, A. V., Parinov, I. A. and Grigoryan, M. N. (2020). Estimation of the output parameters of a numerical model of a cantilever-type piezoelectric generator with attached mass and active termination upon pulsed excitation. *IOP Conference Series: Material Science Engineering*, 913, 022014.

328. Preumont, A. (2018). *Vibration Control of Active Structures* (Springer International Publishing AG).

329. Qu, Y., Jin, F. and Yang, J. (2021). Magnetically induced charge redistribution in the bending of a composite beam with flexoelectric semiconductor and piezomagnetic dielectric layers. *Journal of Applied Physics*, 129(6), 064503.

330. Qu, Y., Jin, F. and Yang, J. (2022). Vibrating flexoelectric microbeams as angular rate sensors. *Micromachines*, 13(8), 1243.

331. Rajeswari, N. R. and Malliga, P. (2015). Analytical approach for optimization design of MEMS based microneedles in drug delivery system. *Journal of Mechanical Science and Technology*, 29, 3405–3415.

332. Ramlan, R., Brennan, M. J., Mace, B. R. and Kovacic, I. (2010). Potential benefits of a non-linear stiffness in an energy harvesting device. *Nonlinear Dyn.* 59, 545–558.

333. Rastehkenari, S. F. and Ghadiri, M. (2021). Nonlinear random vibrations of functionally graded porous nanobeams using equivalent linearization method. *Applied Mathematical Modelling*, 89, 1847–1859.

334. Raymond, M. V. and Smyth, D. M. (1995). Non-stoichiometry defects and charge transport in PZT. In *Science and Technology of Electroceramic Thin Films*, eds. Auciello, O. and Waser, R. (Kluwer Academic Publishers, Dordrecht, The Netherlands) pp. 315–325.

335. Reaney, I. M. and Damjanovic, D. (1996). Crystal structure and domain-wall contributions to the piezoelectric properties of strontium bismuth titanate ceramics. *Journal of Applied Physics*, 80, 4223–4225.

336. Reaney, I. M. and Ubic, R. (1999). Dielectric and structural characteristics of perovskites and related materials as a function of tolerance factor. *Ferroelectrics*, 228, 23–38.

337. Rezaei-Hosseinabadi, N., Tabesh, A., Dehghani, R. and Aghili, A. (2015). An efficient piezoelectric windmill topology for energy harvesting from low-speed air flows. *IEEE Transactions on Industrial Electronics*, 62, 3576–3583.

338. Ross, B. and Samko, S. G. (1995). Fractional integration operator of a variable order in the Hölder spaces. *International Journal of Mathematics and Mathematical Sciences*, 18, 777–788.

339. Rubin, B. S. (1974). Fractional integrals in Hölder spaces with weight and potential type operators. *News of AS ArmSSR. Mathematics*, 9(4), 308–324 (in Russian).

340. Ryu, H. and Kim, S. W. (2021). Emerging pyroelectric nanogenerators to convert thermal energy into electrical energy. *Small*, 17(9), 1903469.

341. Safaei, M., Sodano, H. A. and Anton, S. R. (2019). A review of energy harvesting using piezoelectric materials: State-of-the-art a decade later (2008–2018). *Smart Materials and Structures*, 28, 113001.
342. Salah, M., Boukhoulda, F. B., Nouari, M. and Bendine, K. (2020). Temperature variation effect on the active vibration control of smart composite beam. *Acta Mechanica et Automatica*, 14(3), 166–174.
343. Samko, N., Samko, S. and Vakulov, B. (2010). Fractional integrals and hypersingular integrals in variable order Hölder spaces on homogeneous spaces. *Journal of Function Spaces and Applications*, 8(3), 215–244.
344. Samko, N. and Vakulov, B. (2011). Spherical fractional and hypersingular integrals of variable order in generalized Hölder spaces with variable characteristic. *Mathematische Nachrichten*, 284, 355–369.
345. Samko, S. G. (1977). Generalized Riesz potentials and hypersingular integrals, their symbols and inversion. *Doklady Akademii Nauk SSSR*, 232, 528–531 (in Russian).
346. Samko, S. G. (1995). Fractional integration and differentiation of variable order. *Analysis Mathematica*, 21, 213–236.
347. Samko, S. G. (1998). Differentiation and integration of variable order and the spaces. *Contemporary Mathematics*, 212, 203–219.
348. Samko, S. G. (2002). *Hypersingular Integrals and Their Applications* (Taylor and Frances, London).
349. Samko, S. G., Marichev, A. A. and Kilbas, O. I. (1993). *Fractional Integrals and Derivatives: Theory and Applications* (Gordon and Breach Science Publishers, Philadelphia, PA).
350. Samko, S. G. and Ross, B. (1993). Integration and differentiation to a variable fractional order. *Integral Transforms and Special Functions*, 1, 277–300.
351. Samko, S. G. and Vakulov, B. G. (2000). On equivalent norms in fractional order function spaces of continuous functions on the unit sphere. *Fractional Calculus and Applied Analysis*, 3(4), 401–433.
352. Sarker, M. R., Julai, S., Sabri, M. F. M., Said, S. M., Islam, M. M. and Tahir, M. (2019). Review of piezoelectric energy harvesting system and application of optimization techniques to enhance the performance of the harvesting system. *Sensors and Actuators A: Physical*, 300, 111634.
353. Scherera, R., Kallab, S. L., Tangc, Y. and Huang, J. (2011). The Grünwald-Letnikov method for fractional differential equations. *Computers & Mathematics with Applications*, 62, 902–917.
354. Shaikh, F. K. and Zeadally, S. (2016). Energy harvesting in wireless sensor networks: A comprehensive review. *Renewable and Sustainable Reviews*, 55, 1041–1054.

355. Shao, Y., Xu, M., Shao, S. and Song, S. (2020). Effective dynamical model for piezoelectric stick-slip actuators in bi-directional motion. *Mechanical Systems and Signal Processing*, 145, 106964.
356. Sheeraz, M. A., Malik, M. S., Rahman, K., Elahi, H., Khurram, M., Eugeni, M. and Gaudenzi, P. (2022). Multimodal piezoelectric wind energy harvester for aerospace applications. *International Journal of Energy Research*, 46(10), 13698–13710.
357. Shevtsov, S., Akopyan, V., Rozhkov, E., Chebanenko, V., Yang, C.-C., Lee, C.-Y. J. and Kuo, C.-X. (2016). Optimization of the electric power harvesting system based on the piezoelectric stack transducer. In *Advanced Materials — Manufacturing, Physics, Mechanics and Applications*, Vol. 175, eds. Parinov, I., Chang S. H., Topolov, V. (Springer Cham, Heidelberg, New York, Dordrecht, London) pp. 639–650.
358. Shevtsov, S., Soloviev, A. N., Parinov, I. A., Cherpakov, A. V. and Chebanenko, V. A. (2018). *Piezoelectric Actuators and Generators for Energy Harvesting - Research and Development* (Springer Cham, Switzerland).
359. Shimakawa, Y., Kubo, Y., Nakagawa, Y., Goto, S., Kamiyama, T., Asano, H. and Izumi, F. (2001). Crystal structure and ferroelectric properties of $A\text{Bi}_2\text{Ta}_2\text{O}_9$ (A = Ca, Sr, and Ba). *Physical Review B*, 61, 6559–6554.
360. Shimakawa, Y., Kubo, Y., Nakagawa, Y., Kamiyama, T., Asano, H. and Izumi, F. (1999). Crystal structures and ferroelectric properties of $\text{SrBi}_2\text{Ta}_2\text{O}_9$ and $\text{Sr}_{0.8}\text{Bi}_{2.2}\text{Ta}_2\text{O}_9$. *Applied Physics Letters*, 74, 1904–1906.
361. Shulman, H. S., Testorf, M., Damjanovic, D. and Setter, N. (1996). Microstructure, electrical conductivity and piezoelectric properties of bismuth titanate. *Journal of the American Ceramic Society*, 79, 3124–3128.
362. Shuvaev, A. T., Vlasenko, V. G., Drannikov, D. S. and Zarubin, I. A. (2005). "Structure and properties of $\text{Bi}_4\text{Pb}_{1.5}\text{Ti}_{4.5}\text{O}_{16.5}$ and $\text{Bi}_5\text{Ca}_{0.5}\text{GaTi}_{3.5}\text{O}_{16.5}$. *Inorganic Materials*, 41, 1085–1088.
363. Skaliukh, A. S. (1997). Oscillations of transversely polarized rod transducers with partially electroded ends. In *Integro-differential Operators and Their Applications* (Rostov State University Press, Rostov-on-Don), Vol. 2, pp. 133–138 (in Russian).
364. Skaliukh, A. S. (2019). Functional dependence of physical characteristics on irreversible parameters during electromechanical action on ferroelectric ceramics. *Tomsk State University Bulletin, Mathematics and Mechanics*, 58, 128–141 (in Russian).

365. Skaliukh, A. S., Gerasimenko, T. E., Oganesyan, P. A., Soloviev, A. N. and Solovieva, A. A. (2019). Geometric and physical parameters influence on the resonant frequencies of ultrasonic vibrations of a medical scalpel. In *Advanced materials: Proceedings of the International Conference on Physics and Mechanics of New Materials and Their Applications*, PHENMA 2018, Vol. 224, eds. Parinov, I. A., Chang, S.-H. and Kim, Y.-H. (Springer International Publishing, USA) pp. 507–521.

366. Sladek, J., Sladek, V., Xu, M. and Deng, Q. (2021). A cantilever beam analysis with flexomagnetic effect. *Meccanica*, 56(9), 2281–2292.

367. Smolensky, G. A., Bokov, V. A., Isupov, V. A.. *et al.* (1985). *Physics of Ferroelectric Phenomena* (Nauka, Leningrad) (in Russian).

368. Smolensky, G. A., Isupov, V. A. and Agranovskaya, A. I. (1961). Ferroelectrics of the oxygen-octahedral type with layered structure. *Soviet Physics, Solid State*, 3, 651–655.

369. Smolenskii, G. A. and Kozhevnikova, N. V. (1951). On the question of the emergence of ferroelectricity. *Doklady Akademii Nauk SSSR*, 76, 519–522 (in Russian).

370. Snedden, A., Charkin, D. O., Dolgikh, V. A. and Lightfoot, P. (2005). Crystal structure of the 'mixed-layer' Aurivillius phase $Bi_5TiNbWO_{15}$. *Journal of Solid State Chemistry*, 178, 180–184.

371. Snedden, A., Hervoches, C. H. and Lightfoot, P. (2003). Ferroelectric phase transitions in $SrBi_2Nb_2O_9$ and $Bi_5Ti_3FeO_{15}$: A powder neutron diffraction study. *Physical Review B*, 67, 092102.

372. Soloviev, A. N. (2005). Direct and inverse problems for finite elastic and electroelastic bodies. DrSc Thesis (Physics and Mathematics), Rostov State University Press, Rostov-on-Don, p. 296. (in Russian).

373. Soloviev, A. N., Chebanenko, V. A., Oganesyan, P. A., Chao, S. F. and Liu, Y. M. (2019). Applied theory for electro-elastic plates with non-homogeneous polarization. *Materials Physics & Mechanics*, 42(2), 242–255.

374. Soloviev, A. N., Chebanenko, V. A. and Parinov, I. A. (2018). Mathematical modelling of piezoelectric generators on the base of the Kantorovich method. In *Analysis and Modelling of Advanced Structures and Smart Systems*, Vol. 81, eds. Altenbach, H., Carrera, E. and Kulikov, G. (Springer, Singapore) pp. 227–258.

375. Soloviev, A. N., Chebanenko, V. A., Parinov, I. A. and Oganesyan, P. A. (2019). Applied theory of bending vibrations of a piezoelectric bimorph with a quadratic electric potential distribution. *Materials Physics and Mechanics*, 42, 65–73.

376. Soloviev, A. N., Chebanenko, V. A., Zhilyaev, I. V., Cherpakov, A. V. and Parinov, I. A. (2020). Numerical optimization of the cantilever piezoelectric generator. *Materials Physics and Mechanics*, 44, 94–102.

377. Soloviev, A. N., Do, B. T., Chebanenko, V. A. and Parinov, I. A. (2022). Flexural vibrations of a composite piezoactive bimorph in an alternating magnetic field: Applied theory and finite-element simulation. *Mechanics of Composite Materials*, 58(4), 471–482.

378. Soloviev, A. N., Epikhin, A. N., Glushko, N. I., Lesnyak, O. N. and Solovieva. A. A. (2021). Finite element modeling of a lensotome tip with piezoelectric drive. In *Proceedings of the 2019 International Conference on "Physics, Mechanics of New Materials and Their Applications", PHENMA 2020*, Vol. 10, eds. Parinov, I. A., Chang, S. H., Kim, Y. H. and Noda, N. A. (Springer Nature, Cham, Switzerland) pp. 425–444.

379. Soloviev, A. N., Oganesyan, P. A. and Fomenko, E. I. (2023). Investigation of the efficiency of a shear piezoelectric generator using porous piezoceramics. In *Physics and Mechanics of New Materials and Their Applications: Proceedings of the International Conference PHENMA 2021–2022*, Vol. 20, eds. Parinov, I. A., Chang, S. H. and Soloviev, A. N. (Springer International Publishing, USA), pp. 429–435.

380. Soloviev, A. N., Oganesyan, P. A., Lupeiko, T. G., Kirillova, E. V., Chang, S.-H. and Yang, C.-D. (2016). Modeling of non-uniform polarization for multi-layered piezoelectric transducer for energy harvesting devices. In *Advanced Materials — Manufacturing, Physics, Mechanics and Applications*. 175, eds. Parinov, I. A., Chang, S. H. and Topolov, V. Y. (Springer Cham, Heidelberg, New York, Dordrecht, London) pp. 651–658.

381. Soloviev, A. N., Parinov, I. A. and Cherpakov, A. V. (2021). Modeling the cantilever type PEG with proof mass and active pinching by using the porous piezoceramics with effective properties. In *Proceedings of the 2019 International Conference on "Physics, Mechanics of New Materials and Their Applications", PHENMA 2020*, Vol. 10, eds. Parinov, I. A., Chang, S. H., Kim, Y. H. and Noda, N. A. (Springer Nature, Cham, Switzerland) pp. 481–493.

382. Soloviev, A. N., Parinov, I. A., Cherpakov, A. V. and Chebanenko, V. A. (2021). Investigation of the output parameters of a cantilever PEG with two piezoelectric elements at vibration excitation by rotating drive. In *Proceedings of the 2nd Annual International Conference on Material, Machines and Methods for Sustainable Development (MMMS2020)*, (Springer International Publishing, USA), pp. 705–709.

383. Soloviev, A. N., Parinov, I. A., Cherpakov, A. V., Chebanenko, V. A. and Rozhkov, E. V. (2018). Analysing the output characteristics of a double-console PEG based on numerical simulation. *Materials Physics and Mechanics*, 37, 168–175.

384. Soloviev, A. N., Parinov, I. A., Cherpakov, A. V., Chebanenko, V. A., Rozhkov, E. V. and Duong, L. V. (2018). Analysis of the performance of the cantilever type piezoelectric generator based on finite element modeling. In *Advances in Structural Integrity*, eds. Prakash, R., Jayaram, V. and Saxena, A. (Springer Nature, Singapore) pp. 291–301.

385. Soloviev, A. N. and Vernigora, G. D. (2010). Identification of effective properties of the piezocomposites on the base of FEM modeling with ACELAN. In *Piezoceramic Materials and Devices*, ed. Parinov, I. (Nova Science Publishers, New York) pp. 219–242.

386. Solovyev, A. N. and Duong, L. V. (2016). Optimization for the harvesting structure of the piezoelectric bimorph energy harvesters of circular plate by reduced order finite element analysis. *International Journal of Applied Mechanics*, 8, 1–17.

387. Sridaranea, R., Kalaiselvi, B. J., Akila, B., Subramanian, S. and Murugan, R. (2005). Dielectric properties of $Sr_{1+x}Bi_{2-(2/3)x}$ $(V_xTa_{1-x})_2O_9$ [$x = 0, 0.1 , 0.2$] ceramics. *Physica B*, 357, 439–444.

388. Srinivas, A., Boey, F., Sritharan, T., Dong, W. K. and Kug, S. H. (2004). Processing and study of dielectric and ferroelectric nature of $BiFeO_3$ modified $SrBi_2Nb_2O_9$. *Ceramic Internationals*, 30, 1427–1430.

389. Srinivas, A., Boey, F., Sritharan, T., Dong, W. K. and Kug, S. H. (2004). Study of piezoelectric, magnetic and magnetoelectric measurements on $SrBi_3Nb_2FeO_{12}$ ceramic. *Ceramic Internationals*, 30, 1431–1433.

390. Srinivas, A., Kim, D.-W., Hong, K. S. and Suryanarayana, S. V. (2004). Study of magnetic and magnetoelectric measurements in bismuth iron titanate ceramic $Bi_8Fe_4Ti_3O_{24}$. *Materials Research Bulletin*, 39, 55–61.

391. Srinivas, A., Kumar, M. M. and Suryanarayana, S. V. (1999). Investigation of dielectric and magnetic nature of $Bi_7Fe_3Ti_3O_{21}$. *Materials Research Bulletin*, 34, 989–996.

392. Srinivas, A., Suryanarayana, S. V., Kumar, G. S. and Kumar, M. (1999). Magnetoelectric measurements on $Bi_5FeTi_3O_{15}$ and Bi_6Fe_2 Ti_3O_{18}. *Journal of Physics: Condensed Matter*, 11, 3335–3340.

393. Stanton, S. C., Mann, B. P. and Owens, B. A. M. (2012). Melnikov theoretic methods for characterizing the dynamics of a bistable

piezoelectric inertial generator in complex spectral environments. *Physica D*, 241, 711–720.

394. Stanton, S. C., McGehee, C. C. and Mann, B. P. (2010). Nonlinear dynamics for broadband energy harvesting: Investigation of a bistable piezoelectric inertial generator. *Physica D: Nonlinear Phenomena*, 239, 640–653.

395. Su, Y. and Zhou, Z. (2020). Electromechanical analysis of flexoelectric nanosensors based on nonlocal elasticity theory. *Micromachines*, 11(12), 1077.

396. Suarez, D. Y., Reaney, I. M. and Lee, W. E. (2001). Relation between tolerance factor and T_C in Aurivillius compounds. *Journal of Materials Research*, 16, 3139–3149.

397. Subbarao, E. C. (1961). Ferroelectricity in $Bi_4Ti_3O_{12}$ and its solid solutions. *Physical Review Journals*, 122, 804–807.

398. Subbarao, E. C. (1961). Ferroelectricity in mixed bismuth oxides with layered-type structure. *The Journal of Chemical Physics*, 34, 695–696.

399. Subbarao, E. C. (1962). A family of ferroelectric bismuth compounds. *Journal of Physics and Chemistry of Solids*, 23, 665–676.

400. Subbarao, E. C. (1962). Crystal chemistry of mixed bismuth oxides with layer-type structure. *Journal of the American Ceramic Society*, 45, 166–169.

401. Sundaram, N. G. and Row, G. N. G. (2002). Structure determination at room temperature and phase transition studies above T_C in $ABi_4Ti_4O_{15}$ (A = Ba, Sr or Pb). *Bulletin of Materials Science*, 25, 275–281.

402. Suryanarayana, S. V. (1994). Magnetoelectric interaction phenomena in materials. *Bulletin of Materials Science*, 17, 1259–1270.

403. Suzuki, M. (1995). Doping effect in layer structured ferroelectrics $SrBi_2Nb_2O_9$. *Journal of the Ceramic Society of Japan*, 103, 1088–1090.

404. Syta, A., Litak, G., Lenci, S. and Scheffler, M. (2014). Chaotic vibrations of the duffing system with fractional damping. *Chaos*, 24, 013107.

405. Tagantsev, A. K. and Yudin, P. V. (eds.). (2016). *Flexoelectricity in Solids: From Theory to Applications* (World Scientific, Singapore).

406. Takenaka, T. and Sakata, K. (1984). Grain orientation effects on electrical properties of bismuth layer-structured ferroelectric $Pb_{(1-x)}(NaCe)_{x/2}Bi_4Ti_4O_{15}$ solid solution. *Journal of Applied Physics*, 55, 1092–1099.

407. Talebi, S., Arvin, H. and Beni, Y. T. (2023). Thermal free vibration examination of sandwich piezoelectric agglomerated randomly oriented CNTRC Timoshenko beams regarding pyroelectricity. *Engineering Analysis with Boundary Elements*, 146, 500–516.

408. Tang, L., Wu, H. and Yang, Y. (2011). Optimal performance of nonlinear energy harvesters. *Proceedings of ICAST2011: 22nd International Conference on Adaptive Structures and Technologies*, ICAST 2011, Corfu, Greece, October 10–12, 2011.

409. Tang, L., Yang, Y. and Soh, C. K. (2010). Toward broadband vibration-based energy harvesting. *Journal of Intelligent Material Systems and Structures*, 21, 1867–1897.

410. Tang L., Yang Y. and Soh C. K. (2012). Improving functionality of vibration energy harvesters using magnets. *Journal of Intelligent Material Systems and Structures*, 23, 1433–1449.

411. Tang, L., Yang, Y. and Soh, C. K. (2013). Broadband vibration energy harvesting techniques. In *Advances in Energy Harvesting Methods*, eds. Elvin, N. and Erturk, A. (Springer, Berlin, Heidelberg, Germany) pp. 17–61.

412. Tarasov, V. E. (2010). *Fractional Dynamics: Applications of Fractional Calculus to Dynamics of Particles, Fields and Media* (Springer, Berlin, Germany).

413. Tellier, J., Boullay, Ph., Créon, N. and Mercurio, D. (2005). The crystal structure of the mixed-layer Aurivillius phase $Bi_5Ti_{1.5}W_{1.5}O_{15}$. *Solid State Sciences*, 7, 1025–1034.

414. Tellier, J., Boullay, Ph., Manier, M. and Mercurio, D. (2004). A comparative study of the Aurivillius phase ferroelectrics $CaBi_4Ti_4O_{15}$ and $BaBi_4Ti_4O_{15}$. *Journal of Solid State Chemistry*, 177, 1829–1837.

415. Thai, T. Q., Zhuang, X. and Rabczuk, T. (2023). Curved flexoelectric and piezoelectric micro-beams for nonlinear vibration analysis of energy harvesting. *International Journal of Solids and Structures*, 264, 112096.

416. Tho, N. C., Thanh, N. T., Tho, T. D., Van Minh, P. and Hoa, L. K. (2021). Modelling of the flexoelectric effect on rotating nanobeams with geometrical imperfection. *Journal of the Brazilian Society of Mechanical Sciences and Engineering*, 43, 1–22.

417. Tianchen, Y., Jian, Y., Ruigang, S. and Xiaowei, L. (2014). Vibration energy harvesting system for railroad safety based on running vehicles. *Smart Materials and Struct*ures, 23, 125046.

418. Tommasino, D., Moro, F., Bernay, B., De Lumley Woodyear, T., de Pablo Corona, E. and Doria, A. (2022). Vibration energy harvesting by means of piezoelectric patches: Application to aircrafts. *Sensors*, 22(1), 363.

419. Uchino, K. (2014). Piezoelectric actuator renaissance. *Energy Harvesting and Systems*, 1, 45–56.

420. Vaghefpour, H. (2021). Nonlinear vibration and tip tracking of cantilever flexoelectric nanoactuators. *Iranian Journal of Science and Technology, Transactions of Mechanical Engineering*, 45(4), 879–889.

421. Vaghefpour, H. and Arvin, H. (2019). Nonlinear free vibration analysis of pre-actuated isotropic piezoelectric cantilever Nano-beams. *Microsystem Technologies*. 25, 4097–4110.

422. Vakulov, B. G. (1986). Potential type operator on a sphere in generalized Hölder classes. *Soviet Mathematics (Iz. VUZ)*, 30(11), 90–94.

423. Vakulov, B. G. (2001). On the action of the Riesz potential operator of complex order on R_n in Hölder weight spaces. *News of Universities of the North-Caucasus Region. Natural Sciences*, 4, 47–49 (in Russian).

424. Vakulov, B. G. (2005). Spherical potential-type operators in Hölder weight spaces of variable order. *Vladikavkaz Mathematical Journal*, 7(2), 26–40 (in Russian).

425. Vakulov, B. G. (2006). Complex-order spherical potentials in spaces of variable generalized Hölder order. *Doklady Mathematics*, 73(2), 165–168.

426. Vakulov, B. G. and Drobotov, Yu. E. (2020). Riesz potential with summable density in spaces of variable Hölder order. *Mathematical Notes*, 108(5), 652–660.

427. Vakulov, B. G. and Karapetyants, N. K. (2003). Potential type operators on a sphere with singularities at the poles. *Doklady RAN*, 392(2), 151–154 (in Russian).

428. Vakulov, B. G., Karapetyants, N. K. and Shankishvili, L. D. (2002). Spherical potentials of complex order in generalized Hölder spaces with weight. *Doklady Mathematics*, 65(1), 35–38.

429. Vakulov, B. G., Karapetyants, N. K. and Shankishvili, L. D. (2003). Spherical convolution operators with a power-logarithmic kernel in generalized Hölder spaces. *News of Universities. Mathematics*, 2, 3–14 (in Russian).

430. Vakulov, B. G. and Kochurov, E. S. (2010). Operators of fractional integration and differentiation of a variable order in the Hölder spaces. *Vladikavkaz Mathematical Journal*, 12(4), 3–11 (in Russian).

431. Vakulov, B. G. and Kochurov, E. S. (2011). Zygmund-type estimates for operators of fractional integration and differentiation of a variable order. *News of Universities of the North-Caucasus Region. Natural Sciences*. Special issue, 15–17 (in Russian).

432. Vakulov, B. G., Kochurov, E. S. and Samko, N. G. (2011). Russian Mathematics. *News of Universities. Mathematics*, 55(6), 20–28 (in Russian).
433. Vakulov, B. G., Kostetskaya, G. S. and Drobotov, Y. E. (2018). Riesz potential in generalized Hölder spaces. In *Fractal Approaches for Modeling Financial Assets and Predicting Crises*, eds. Nekrasova, I., Karnaukhova, O. and Christiansen B. (IGI Global, Pennsylvania, USA), pp. 249–273.
434. Vakulov, B. G. and Samko, S. G. (1987). On equivalent normalizations in spaces of fractional smoothness functions on a sphere. *News of Universities. Mathematics*, 12, 68–71 (in Russian).
435. Van Tuyen, B. (2022). Free vibration behaviors of nanoplates resting on viscoelastic medium. *Arabian Journal for Science and Engineering*, 48, 11511–11524.
436. Villegas, M., Caballero, A. C., Moure, C., Duran, P. and Fernandez, J. F. (1999). Low temperature sintering and electrical properties of chemically W-doped $Bi_4Ti_3O_{12}$ ceramics. *Journal of the European Ceramic Society*, 19, 1183–1186.
437. Vlasenko, V. G., Shuvaev, A. T. and Drannikov, D. S. (2005). A structural study of the Aurivillius phases by X-ray powder diffraction. *Powder Diffraction*, 20(1), 1–6.
438. Vlasenko, V. G., Shuvaev, A. T., Zarubin, I. A., Shuvaeva, E. T. and Petin G. P. (2003). Crystal structure of new Aurivillius phases. *Research in Russia*, 55, 654–663.
439. Vlasenko, V. G., Shuvaeva, V. A., Levchenkov, S. I., Zubavichus, Y. V. and Zubkov, S. V. (2014). Structural, electrical and magnetic characterisation of a new Aurivillius phase $Bi_{5-x}Th_xFe_{1+x}Ti_{3-x}O_{15}$ ($x = 1/3$). *Journal of Alloys and Compounds*, 610, 184–188.
440. Vlasenko, V. G., Zubkov, S. V. and Shuvaeva, V. A. (2015). Structure and dielectric properties of solid solutions $Bi_7Ti_{4+x}W_xNb_{1-2x}O_{21}$ ($x = 0$–0.5). *Physics of the Solid State*, 57, 900–906.
441. Vlasenko, V. G., Zubkov, S. V., Shuvaeva, V. A., Abdulvakhidov, K. G. and Shevtsova, S. I. (2014). Crystal structure and dielectric properties of Aurivillius phases $A_{0.5}Bi_{4.5}B_{0.5}Ti_{3.5}O_{15}$ (A = Na, Ca, Sr, Pb; B = Cr, Co, Ni, Fe, Mn, Ga). *Physics of the Solid State*, 56, 1554–1560.
442. Wang, B., Gu, Y., Zhang, S. and Chen, L. Q. (2019). Flexoelectricity in solids: Progress, challenges, and perspectives. *Progress in Materials Science*, 106, 100570.
443. Wang, B. and Li, X. F. (2021). Flexoelectric effects on the natural frequencies for free vibration of piezoelectric nanoplates. *Journal of Applied Physics*, 129(3), 034102.

444. Wang, C., Lai, S.-K., Wang, J.-M., Feng, J.-J. and Ni, Y.-Q. (2021). An ultra-low-frequency, broadband and multi-stable tri-hybrid energy harvester for enabling the next-generation sustainable power. *Applied Energy*, 291, 116825.

445. Wang, C., Zhang, C. and Wang, W. (2017). Low-frequency wide-band vibration energy harvesting by using frequency up-conversion and quin-stable nonlinearity. *Journal of Sound and Vibration*, 399, 169–181.

446. Wang, C., Zhang, Q., Wang, W. and Feng, J. (2018). A low-frequency, wideband quad-stable energy harvester using combined nonlinearity and frequency up-conversion by cantilever-surface contact. *Mechanical Systems and Signal Processing*, 112, 305–318.

447. Wang, G., Liao, W.-H., Yang, B., Wang, X., Xu, W. and Li, X. (2018). Dynamic and energetic characteristics of a bistable piezoelectric vibration energy harvester with an elastic magnifier. *Mechanical Systems and Signal Processing*, 105, 427–446.

448. Wang, H., Jasim, A. and Chen, X. (2018). Energy harvesting technologies in roadway and bridge for different applications — A comprehensive review. *Applied Energy*, 212, 1083–1094.

449. Wang, J., Shi, Z. and Han, Z. (2013). Analytical solution of piezo-electric composite stack transducers. *Journal of Intelligent Material Systems and Structures*, 24, 1626–1636.

450. Wang, L., Yu, P., Zhang, S., Zhao, Z. and Jin, J. (2023). Electrome-chanical coupling model of variable-section piezoelectric composite beams in longitudinal vibration. *International Journal of Mechanical Sciences*, 241, 107973.

451. Wang, L., Zhao, L., Jiang, Z., Luo, G., Yang, P., Han, X. and Maeda, R. (2019). High accuracy comsol simulation method of bimorph cantilever for piezoelectric vibration energy harvesting. *AIP Advances*, 9(9), 095067.

452. Wang, Q., Bowen, C. R., Lewis, R., Chen, J., Lei, W., Zhang, H., Li, M.-Y. and Jiang, S. (2019). Hexagonal boron nitride nanosheets doped pyroelectric ceramic composite for high-performance thermal energy harvesting. *Nano Energy*, 60, 144–152.

453. Wang, S., Yang, Z.; Kan, J., Chen, S., Chai, C. and Zhang, Z. (2021). Design and characterization of an amplitude-limiting rotational piezoelectric energy harvester excited by a radially dragged magnetic force. *Renewable Energy*, 177, 1382–1393.

454. Wang, W., Cao, J., Mallick, D., Roy, S. and Lin, J. (2018). Comparison of harmonic balance and multi-scale method in characterizing the response of monostable energy harvesters. *Mechanical Systems and Signal Processing*, 108, 252–261.

455. Wang, W., Li, M., Jin, F., He, T. and Ma, Y. (2023). Nonlinear magnetic-mechanical-thermo-electric coupling characteristic analysis on the coupled extensional and flexural vibration of flexoelectric energy nanoharvester with surface effect. *Composite Structures*, 116687.

456. Waser, R. (1991). Bulk conductivity and defect chemistry of acceptor-doped strontium titanate in the quenched state. *Journal of the American Ceramic Society*, 74, 1934–1940.

457. Wei, C. and Jing, X. (2017). A comprehensive review on vibration energy harvesting: Modelling and realization. *Renewable and Sustainable Energy Reviews*, 74, 1–18.

458. Werens, M. (1981). Best approximation on the unit sphere in R_k. In *Proceedings of the Conference held at the Mathematical Research Institute at Oberwolfach Black Forest, August 9–16, 1980*, eds. Butzer, P., Gärlich, E. and Szökefalvi-Nagy B. (Birkhäuser, Basel, Switzerland), pp. 233–245.

459. Weyl, H. (1917). Bemerkugen zum begriff des differentialquotienten gebrochener ordung. *Vierteljahrcsschrift der Naturforschenden Gesellschaft in Zurich*, 62(1–2), 296–302 (in German).

460. Wiggins, S. (2003). *Introduction to Applied Nonlinear Dynamical Systems and Chaos* (Springer, New York).

461. Wischke, M., Masur, M., Goldschmidtboeing, F. and Woias, P. (2010). Electromagnetic vibration harvester with piezoelectrically tunable resonance frequency. *Journal of Micromechanics and Microengineering*, 20, 035025.

462. Withers, R. L., Thompson, J. G. and Rae, A. D. (1991). The crystal chemistry underlying ferroelectricity in $Bi_4Ti_3O_{12}$, Bi_3TiNbO_9 and Bi_2WO_6. *Journal of Solid State Chemistry*, 94, 404–417.

463. Won, S. S., Seo, H., Kawahara, M., Glinsek, S., Lee, J., Kim, Y., Jeong, C. K., Kingon, A. I. and Kim, S.-H. (2019). Flexible vibrational energy harvesting devices using strain-engineered perovskite piezoelectric thin films. *Nano Energy*, 55, 182–192.

464. Wong-Ng, W., Huang, Q., Cook, L. P., Levin, I., Kaduk, J. A., Mighell, A. D. and Suh, J. (1950). Crystal chemistry and crystallography of the Aurivillius phase $Bi_5AgNb_4O_{18}$. *Journal of Solid State Chemistry*, 177, 3359–3367.

465. Wu, N., Bao, B. and Wang, Q. (2021). Review on engineering structural designs for efficient piezoelectric energy harvesting to obtain high power output. *Engineering Structures*, 235, 112068.

466. Wu, N., Wang, Q. and Xie, X. (2013). Wind energy harvesting with a piezoelectric harvester. *Smart Materials and Structures*, 22, 095023.

467. Wu, Y. and Cao G. Z. (1999). Enhanced ferroelectric properties and lowered bismuth niobates with vanadium doping. *Applied Physics Letters*, 75, 2650–2652.

468. Wu, Y. and Cao G. Z. (2000). Ferroelectric and dielectric properties of strontium bismuth niobate vanadates. *Journal of Materials Research*, 15, 1583–1590.

469. Wu, Y. and Cao G. Z. (2000). Influence of vanadium doping on ferroelectric properties of strontium bismuth niobates. *Journal of Materials Science Letters*, 19, 267–269.

470. Wu, Y., Forbess, M., Seraji, S., Limmer, S. J., Chou, T. P. and Cao, G. Z. (2001). Impedance study of strontium bismuth tantalate vanadate ferroelectrics. *Materials Science and Engineering*, 86, 70–78.

471. Wu, Y., Forbess, M. J., Seraji, S., Limmer, S. J., Chou, T. P., Nguyen, C. and Cao, G. (2001). Doping effect in layer structured $SrBi_2Nb_2O_9$ ferroelectrics. *Journal of Applied Physics*, 90, 5296–5302.

472. Wu, Y., Seraji, S., Forbess, M. J., Limmer, S. J., Chou, T. and Cao, G. Z. (2001). Oxygen-vacancy-induced dielectric relaxation in $SrBi_2(Ta_{0.9}V_{0.1})_2O_9$ ferroelectrics. *Journal of Applied Physics*, 89, 5647–5652.

473. Xiong, L., Tang, L. and Mace, B. R. (2018). A comprehensive study of 2:1 internal-resonance-based piezoelectric vibration energy harvesting. *Nonlinear Dynamics*, 91, 1817–1834.

474. Yan, B., Zhou, S. and Litak, G. (2018). Nonlinear analysis of the tristable energy harvester with a resonant circuit for performance enhancement. *International Journal of Bifurcation and Chaos*, 28, 1850092.

475. Yan, H., Zhang, H., Zhang, Z., Ubic, R. and Reece, M. J. (2005). B-site donor and acceptor doped Aurivillius phase Bi_3NbTiO_9 ceramics. *Journal of the European Ceramic Society*, 26, 2785–2792.

476. Yang, H., Wei, Y., Zhang, W., Ai, Y., Ye, Z. and Wang, L. (2021). Development of piezoelectric energy harvester system through optimizing multiple structural parameters. *Sensors*, 21, 2876.

477. Yang, X., Wang, C. and Lai, S. (2020). A magnetic levitation-based tristable hybrid energy harvester for scavenging energy from low-frequency structural vibration. *Engineering Structures*, 221, 110789.

478. Yaseen, M., Khattak, M. A. K., Humayun, M., Usman, M., Shah, S. S., Bibi, S., Hasnain, B. S. U., Ahmad, S. M., Khan, A., Shah, N., Tahir, A. A. and Ullah, H. (2021). A review of supercapacitors: Materials design, modification, and applications. *Energies*, 14, 7779.

479. Yu, W. J., Kim, Y. I., Ha, D. H., Lee, J. H., Park, Y. K., Seong, S. and Hur, N. H. (1999). A new manganese oxide with the Aurivillius structure: $Bi_2Sr_2Nb_2MnO_{12-\delta}$. *Solid State Communications*, 111, 705–709.

480. Zakharov, Y. N., Sakhnenko, V. P., Parinov, I. A., Raevsky, I. P., Bunin, M. A., Chebanenko, V. A. and Kiseleva, L. I. (2020). Possibilities of the practical use of a stationary strain gradient in the interelectrode volume of unpolarized ferroceramic plates. *Journal of Advanced Dielectrics*, 10(01n02), 2060010.

481. Zakharov, Yu. N., Sitalo, Ye. I., Parinov, I. A. and Boldyrev, N. A. (2021). *Pyroelectric, Flexoelectric and Related Effects in Ferroelectrics, Antiferroelectrics, Ferroelectric Relaxors and Multiferroics* (Southern Federal University Press, Rostov-on-Don) (in Russian).

482. Zarepour, M., Hosseini, S. A. H. and Akbarzadeh, A. H. (2019). Geometrically nonlinear analysis of Timoshenko piezoelectric nanobeams with flexoelectricity effect based on Eringen's differential model. *Applied Mathematical Modelling*, 69, 563–582.

483. Zeng, S., Wang, B. L. and Wang, K. F. (2019). Analyses of natural frequency and electromechanical behavior of flexoelectric cylindrical nanoshells under modified couple stress theory. *Journal of Vibration and Control*, 25(3), 559–570.

484. Zhang, G. Y., Qu, Y. L., Gao, X. L. and Jin, F. (2020). A transversely isotropic magneto-electro-elastic Timoshenko beam model incorporating microstructure and foundation effects. *Mechanics of Materials*, 149, 103412.

485. Zhang, H., Yan, H. and Reece, M. J. (2010). Microstructure and electrical properties of Aurivillius phase $(CaBi_2Nb_2O_9)_{1-x}$ $(BaBi_2Nb_2O_9)_x$ solid solution. *Journal of Applied Physics*, 108, 014109.

486. Zhang, M. and Zhou, Z. (2022). Bending and vibration analysis of flexoelectric beam structure on linear elastic substrates. *Micromachines*, 13(6), 915.

487. Zhang, N., Zheng, S. and Chen, D. (2022). Size-dependent static bending, free vibration and buckling analysis of curved flexomagnetic nanobeams. *Meccanica*, 57(7), 1505–1518.

488. Zhang, N., Zheng, S. and Chen, D. (2022). Size-dependent static bending, free vibration and buckling analysis of simply supported flexomagnetic nanoplates. *Journal of the Brazilian Society of Mechanical Sciences and Engineering*, 44(6), 253.

489. Zhang, X. F., Hu, K. M. and Li, H. (2019). Comparison of flexoelectric and piezoelectric ring energy harvester. *Proceedings of the Institution of Mechanical Engineers, Part C: Journal of Mechanical Engineering Science*, 233(11), 3795–3803.

490. Zhao, D., Gan, M., Zhang, C., Wei, J., Liu, S. and Wang, T. (2018). Analysis of broadband characteristics of two degree of freedom bistable piezoelectric energy harvester. *Materials Research Express*, 5, 085704.

491. Zhao, M., Wang, C., Zhong, W., Wang, J. and Chen, H. (2002). Ferroelectric, piezoelectric and pyroelectric properties of Sr_{1+x} $Bi_{4-x}Ti_{4-x}Ta_xO_{15}$ ceramics ($x = 0$–1). *Japan Journal of Applied Physics*, 41, 1455–1458.

492. Zhao, S. and Erturk, A. (2013). Energy harvesting from harmonic and noise excitation of multilayer piezoelectric stacks: Modeling and experiment. *Active and Passive Smart Structures and Integrated Systems*, 8688, 86881Q.

493. Zhao, S. and Erturk, A. (2014). Deterministic and band-limited stochastic energy harvesting from uniaxial excitation of a multilayer piezoelectric stack. *Sensors and Actuators A: Physical*, 214, 58–65.

494. Zhao, X., Zheng, S. and Li, Z. (2020). Effects of porosity and flexo-electricity on static bending and free vibration of AFG piezoelectric nanobeams. *Thin-Walled Structures*, 151, 106754.

495. Zheng, R., Nakano, K., Hu, H., Su, D. and Cartmell, M. P. (2014). An application of stochastic resonance for energy harvesting in a bistable vibrating system. *Journal of Sound and Vibration*, 333, 2568–2587.

496. Zhi, Y., Chen, A., Vilarinho, P. M., Mantas, P. and Baptista, J. L. (1998). Dielectric properties of Bi doped $SrTiO_3$ ceramics in the temperature range 500–800 K. *Journal of Applied Physics*, 83, 4874–4877.

497. Zhou, Q., Kennedy, B. J. and Howard, C. J. (2003). Structural studies of the ferroelectric phase transition in $Bi_4Ti_3O_{12}$. *Chemistry of Materials*, 15, 5025–5028.

498. Zhou, S., Cao, J., Inman, D. J., Lin, J., Liu, S. and Wang, Z. (2014). Broadband tristable energy harvester: Modeling and experiment verification. *Applied Energy*, 133, 33–39.

499. Zhou, S., Cao, J., Lin, J. and Wang, Z. (2014). Exploitation of a tristable nonlinear oscillator for improving broadband vibration energy harvesting. *The European Physical Journal Applied Physics*, 67, 30902.

500. Zhou, S., Cao, J., Litak, G. and Lin, J. (2018). Numerical analysis and experimental verification of broadband tristable energy harvesters. *Technisches Messen*, 85, 521–532.

501. Zhou, S., Cao, J., Wang, W., Liu, S. and Lin, J. (2015). Modeling and experimental verification of doubly nonlinear magnet-coupled piezoelectric energy harvesting from ambient vibration. *Smart Materials and Structures*, 24, 055008.

502. Zhou, Z., Qin, W. and Zhu, P. (2018). Harvesting performance of quad-stable piezoelectric energy harvester: Modeling and experiment. *Mechanical Systems and Signal Processing*, 110, 260–272.

503. Zhu, D., Tudor, M. J. and Beeby, S. P. (2010). Strategies for increasing the operating frequency range of vibration energy harvesters: A review. *Measurement Science and Technology*, 21, 022001.

504. Zhu, Y. and Shang, H. (2022). Global dynamics of the vibrating system of a tristable piezoelectric energy harvester. *Mathematics*, 10, 2894–2907.

505. Zhuang, X., Nguyen, B. H., Nanthakumar, S. S., Tran, T. Q., Alajlan, N. and Rabczuk, T. (2020). Computational modeling of flexoelectricity — A review. *Energies*, 13(6), 1326.

506. Zubko, P., Catalan, G. and Tagantsev, A. K. (2013). Flexoelectric effect in solids. *Annual Review of Materials Research*, 43, 387–421.

507. Zubkov, S. V. (2021). Crystal structure and dielectric properties of layered perovskite-like solid solutions $Bi_{3-x}Gd_xTiTaO_9$ ($x = 0.0, 0.1, 0.2, 0.3$) with high curie temperature. *Journal of Advanced Dielectrics*, 11, 2160016.

508. Zubkov, S. V., Parinov, I. A. and Kuprina, Y. A. (2022). The Structural and dielectric properties of $Bi_{3-x}Nd_xTi_{1.5}W_{0.5}O_9$ ($x = 0.25, 0.5, 0.75, 1.0$). *Electronics*, 11, 277–287.

509. Zubkov, S. V., Parinov, I. A., Kuprina, Y. A. and Nazarenko, A. V. (2022). Structural and dielectric Properties of $Bi_3Ti_{1.5}W_{0.5}O_9$. *Physics of the Solid State*, 64, 652–657.

510. Zubkov S. V., Parinov I. A., Nazarenko A. V. and Kuprina, Y. A. (2022). Crystal structure, microstructure, piezoelectric and dielectric properties of high-temperature piezoceramics $Bi_{3-x}Nd_xTi_{1.5}W_{0.5}O_9$ ($x = 0, 0.1, 0.2$). *Physics of the Solid State*, 64, 1475–1482.

511. Zubkov, S. V., Parinov, I. A., Nazarenko, A. V. and Kuprina, Y. A. (2023). Microstructure, crystal structure, piezoelectric and dielectric properties of piezoceramic $SrBi_2Nb_2O_9$. In *Physics and Mechanics of New Materials and Their Applications — Proceedings of the International Conference PHENMA 2021–2022*, Springer Proceedings in Materials, Vol. 20, eds. Parinov, I. A., Chang, S.-H. and Soloviev, A. N. (Springer Nature, Cham, Switzerland), pp. 163–174.

512. Zubkov, S. V. and Vlasenko, V. G. (2017). Crystal structure and dielectric properties of layered perovskite-like solid solutions $Bi_{3-x}Y_xTiNbO_9$ ($x = 0, 0.1, 0.2, 0.3$) with high Curie temperature. *Physics of the Solid State*, 59, 2325–2330.

513. Zubkov, S. V., Vlasenko, V. G., Shuvaeva, V. A. and Shevtsova S. I. (2016). Structure and dielectric properties of solid solutions $Bi_7Ti_{4+x}W_xTa_{1-2x}O_{21}$ ($x = 0$–0.5). *Physics of the Solid State*, 58, 42–49.

514. Zurbuchen, M. A., Freitas, R. S., Wilson, M. J., Schiffer, P., Roeckerath, M., Schubert, J., Biegalski, M. D., Mehta, G. H., Comstock, D. J., Lee, J. H., Jia, Y. and Scholm, D. G. (2007). Synthesis and characterization of an $n = 6$ Aurivillius phase incorporating magnetically active manganese $Bi_7 (MnTi)_6O_{21}$. *Applied Physics Letters*, 91, 033113.

Index